# ALGONQUIN PARK A PORTRAIT

*the landscape, wildlife and ecology of an iconic Canadian treasure*

West Rose Lake in autumn

# ALGONQUIN PARK A PORTRAIT

*the landscape, wildlife and ecology of
an iconic Canadian treasure*

## JAN RINIK & MARTIN RINIK

FORMAC PUBLISHING COMPANY LIMITED
HALIFAX

*Praise the Lord from the earth,...
fire and hail, snow and frost,
stormy wind fulfilling his
command!
Mountains and all hills,
fruit trees and all cedars!
Wild animals and all cattle,
creeping things and flying birds!*

*Psalm 148:7-10*

*Dedicated to our mom and to the
beloved memory of our dad.*

Yellow-bellied Sapsucker on White Birch, Lookout Trail

© 2014 Jan Rinik and Martin Rinik

All rights reserved. No part of this book may be reproduced or transmitted in any form or by any means, electronic or mechanical, including photocopying, or by any information storage or retrieval system, without permission in writing from the publisher.

Formac Publishing Company Limited recognizes the support of the Province of Nova Scotia through the Department of Communities, Culture and Heritage. We are pleased to work in partnership with the province to develop and promote our culture resources for all Nova Scotians. We acknowledge the financial support of the Government of Canada through the Canada Book Fund for our publishing activities. We acknowledge the support of the Canada Council for the Arts for our publishing program.

Library and Archives Canada Cataloguing in Publication

Rinik, Jan, 1966-, author, photographer
    Algonquin Park : a portrait : the landscape, wildlife, forests, lakes, streams, and ecology of this iconic Canadian treasure / by Jan Rinik & Martin Rinik.

Includes index.
Includes photographs, paintings, illustrations and maps created by the authors.
ISBN 978-1-4595-0312-0 (bound)

1. Natural history--Ontario--Algonquin Provincial Park.
2. Natural history--Ontario--Algonquin Provincial Park--Pictorial works.
3. Landscapes--Ontario--Algonquin Provincial Park--Pictorial works.
4. Animals--Ontario--Algonquin Provincial Park--Pictorial works.
5. Ecology--Ontario--Algonquin Provincial Park--Pictorial works.
6. Nature photography--Ontario--Algonquin Provincial Park.

I. Rinik, Martin, 1964-, author, photographer

II. Title.

QH106.2.O5R55 2014      578.09713'147
C2013-908003-1

Formac Publishing Company Limited
5502 Atlantic Street
Halifax, Nova Scotia,  Canada
B3H 1G4
www.formac.ca

Printed and bound in Canada

# CONTENTS

| | |
|---|---|
| FOREWORD | 6 |
| PREFACE | 20 |
| ACKNOWLEDGEMENTS | 22 |
| INTRODUCTION – *About Algonquin Provincial Park* | 24 |
| SPRING – *Awakening of Life* | 30 |
| SUMMER – *Days of Bounty* | 74 |
| AUTUMN – *Nature's Palette* | 144 |
| WINTER – *Time to Persist* | 184 |
| BIBLIOGRAPHY | 204 |
| INDEX | 210 |

# FOREWORD

I don't need to tell readers how to enjoy this book. The paintings, the photos and the insights all speak for themselves. They will touch a chord with anyone who has stopped to admire a moose by the roadside, heard the howl of a wolf under starry skies or enjoyed a freshly caught trout on the shore of a wild, far-off lake. For that matter, they will intrigue and fascinate who-knows-how-many people who have not yet made their first personal voyage of discovery to Ontario's most famous park.

But perhaps I can tell you what is particularly impressive and heart-warming about this book for someone who had the privilege of working in Algonquin for almost a whole lifetime. All Algonquin staff members, of course, quite naturally meet and talk to hundreds, if not thousands, of visitors every year. In my day we routinely met fishermen who had been coming to the park for many decades, keen naturalists trying to spot northern specialties like Grey Jay and Spruce Grouse, parents of young families anxious to teach their children about the natural world and senior citizens on whirl-wind bus tours hoping to catch the fall colours at their peak. But then there were the people we hardly ever saw — newcomers to Canada for the most part, people who started their new lives in the big cities down south and, as far as we could tell, seldom ever left. And on the few occasions when we actually did meet new Canadians in Algonquin, they often seemed a little apprehensive about the park environment, or even downright intimidated. 'Is it safe to walk on this trail?' they would ask, 'How can you go into that wilderness and not get lost — and what about the wild animals?' Or when we talked about, say, the fur trade, early logging or the ebb and flow of deer numbers in early Park history — we often had the distinct impression that our new fellow citizens didn't really share our enthusiasm or relate to at least those parts of Canada's natural and cultural heritage.

That is why Jan and Martin's book is a particular stand-

*Moose at Wolf Howl Pond*

out for me. Imagine, two brothers who grew up in far-away Slovakia, on the other side of the world, in an entirely different landscape, who come to Canada with very little English and who almost immediately discover and embrace Algonquin. But they not only fall in love with the park in an aesthetic way, painting and photographing its ever-changing scenery, moods and secrets, they also delve into it intellectually. They try to understand everything they see, they read everything about the park that they can get their hands on and they learn the names of everything they encounter. And then, not content with this already laudable accomplishment, they undertake to write and illustrate this highly personal testament to what they clearly found to be a marvellous place. What's more: when they did this, they had no contract with a publisher or even any idea of how to find one. In other words, they wrote and personally illustrated this book solely out of love for the park, with no real expectation of financial gain. Wow! I was, and still am, very, very impressed.

As I have said to many present and former Algonquin staff, this is far from the only book on Algonquin Park, but, for its combination of aesthetic quality, inclusion of important information, careful emphasis on accuracy and for the unique story of how it came into being, it deserves extra-special recognition. I hope that you, the reader, enjoy it as much as I do.

Dan Strickland

January 2011

Retired Chief Park Naturalist of Algonquin Provincial Park

(1970–2000)

*Costello Creek in autumn*

ALGONQUIN PARK: A PORTRAIT

Whitefish Lake from Centennial Ridges, *painting by Jan Rinik*

# FOREWORD

*Boreal forest at Spruce Bog Trail in Algonquin Park*

# ALGONQUIN PARK: A PORTRAIT

Autumn Splendour from Visitor Centre, *painting by Jan Rinik*

# FOREWORD

*Ring-neck Pond along Highway 60 in Algonquin Provincial Park*

Lake Reflection, *painting by Jan Rinik*

# PREFACE

Beautiful natural destinations have always been highly valued and appreciated by people. Ontario, the most populated province of Canada, can still boast many such wild and beautiful places, and Algonquin Provincial Park is one of them. The exploration of Algonquin's natural treasures is for many people a traditional summer vacation activity, while for others it is an exciting destination all year round. It seems that with our busy lives the value of Algonquin's green forests and rivers, mountains and lakes is growing and gaining significance.

Unique places like Algonquin Provincial Park have also been inspiring to authors. To the sensitive eye of the visual artist, photographer and naturalist it has also been a great source of motivation. The subtle beauty of flowers, the changing autumn colours, the little birds as well as the mighty moose are all reflected in the artwork and photographs throughout the following pages. The paintings and drawings were specifically created for this book and they represent existing places — many of them easily located in the park.

Algonquin Provincial Park is a uniquely beautiful place to visit. For many tourists from all over the world it is a once in a lifetime experience. However, during their trip they may not see all the park's creatures and the beauty around them, or the changing seasons. Whether you are a regular park visitor with a long checklist of species or a first-time tourist, the authors wish to provide you with an artistic, pictorial and relaxing journey through the wildlife and seasons of Algonquin. The book *Algonquin Park: A Portrait* is dedicated to all nature lovers.

Although Algonquin Park is an excellent place to observe wildlife, it is not easy to get close to many shy species of birds or mammals. After one of our successful bird-watching days in 2008, an elderly couple at the Visitor Centre bookstore saw us buying the large and bulky book *Breeding Birds of Ontario*. The gentleman asked a question, obviously a bit doubtful that people could actually see all those birds. Being equipped

# PREFACE

*Yellow-rumped Warbler,* Setophaga coronata, *marks its nesting territory by musical trills in the Algonquin environment*

with binoculars around their necks, he and his wife were presumably birdwatchers, too. We often hear from people visiting Algonquin that they could not see as much as they expected. Our intention in the following pages is to bring the reader closer to the park's creatures by highlighting aspects of their interesting lives. However, a personal experience in wildlife watching is invaluable. It is usually a tough but rewarding effort, not to be replaced by any book.

*Algonquin Park: A Portrait* is a celebration of the wildlife and the seasons of this priceless place, captured by paintbrush, pencil, camera and text. For species illustrated in the following pages, the common English names are accompanied by their scientific names, and they are all included in the index. Please note, however, that this book is not intended to be a reference book or a field guide, and it does not describe 'where and what to watch sites' per se. Please consult the *bibliography* at the end of this book for many fine examples of such books.

# ACKNOWLEDGEMENTS

Of the many people who supported this project, special acknowledgement and recognition is owed to Glenn Barrett, a wildlife biologist and friend. Upon our arrival in Ontario, he directed our attention to Algonquin Park. He also provided background literature as well as helpful comments on earlier drafts of this manuscript.

We are especially indebted to Dan Strickland, a former Chief Naturalist of Algonquin Park. His extensive writings on Algonquin were the source of much of the information conveyed in our book. It was a great privilege that he contributed to our knowledge of the flora and fauna of the park, and willingly provided valuable comments, criticism, suggestions and discussion of an earlier draft to ensure the highest level of objectivity and biological accuracy, especially with the section on White-tailed Deer. We are grateful that he kindly shared his hypothesis shedding light on the changes in abundance of deer in the park, as well as on the complex relationships among deer, wolves and moose.

Several authorities familiar with the park reviewed the text and the book proposal. Many thanks belong to Ron Tozer, a former Algonquin Park Naturalist, for his comments on the text, and especially on birds; Justin Peter, current Senior Park Naturalist at Algonquin Park, for his valuable suggestions; and John Black, past president of the Ontario Field Naturalists, for his corrections and suggestions.

We are thankful to naturalists Richard Aaron, Chris Earley and Troy McMullin, University of Guelph; Greg Thorn, University of Western Ontario; and Lorraine Brown, who all helped with identification of fungi, moss and lichens. Colin Jones, Ontario Ministry of Natural Resources, kindly helped with insect identification. For providing access to ornithological literature, we are thankful to Chip Weseloh, Canadian Wildlife Service.

Our sincere thanks go to George Brylowski for editing the manuscript and for helping with the text composition

## ACKNOWLEDGEMENTS

throughout the whole project. We would also like to express our thanks to Dr. Miroslava Smochko and Dr. Bernard Rubin for their support of this project.

Martin would like to express his sincere personal thanks to his wife, Dana, and to their children Martin and Dominika, for their understanding and support of his work.

*Scenery from Wolf Howl Pond*

# INTRODUCTION:
## *About Algonquin Provincial Park*

Algonquin Provincial Park, located about three hundred kilometres northeast of Toronto and about 250 kilometres northwest of Ottawa, is a treasure chest of Canadian nature. Each year the park attracts thousands of visitors from all over the world. They will all return home with a deep and unforgettable experience of the true Canadian wilderness. Even a one-day trip to the park will let you set aside your busy life and contemplate nature's beauty and harmony.

Algonquin is a large park. Over 7,700 square kilometres (three thousand square miles) of forests are dotted with 2,500 lakes and countless rivers and streams. With its highest point being 585 meters above sea level, it is a realm of woodlands and waterways. The park is accessible by Highway 60, which, for a length of about sixty kilometres (or forty miles), crosses its southwest corner from west to east. There are many other access points, but they're less frequently travelled by visitors.

The name *Algonquin*, for those not familiar with the region, needs a bit of an explanation. In the early 1600s, French explorers entering the St. Lawrence and Ottawa River valleys met native people they came to call *Algonquins*. Unfortunately, the significance of this name was not recorded at the time and remains a mystery. We do know, however, that the original French name referred to many different nations who spoke very similar languages and inhabited a huge area of northeastern North America. This usage passed into English and was still current in 1893 when Algonquin Park was founded. In the words of the Royal Commission that established the park in that year, *'The commissioners suggest that the name of the Reservation be The Algonquin Park, in this way perpetuating the memory of one of the greatest Indian nations that has inhabited the North American continent….They included the Nipissings, Ottawas, Montagnais, Delawares, Wyandots, Mississaugas and over thirty other different tribes. The Nipissings, who are deemed*

# INTRODUCTION

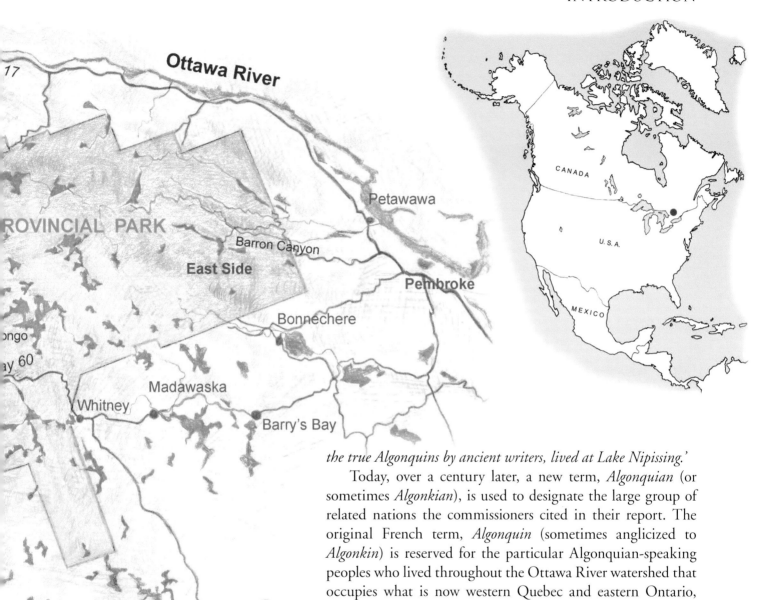

Map of Algonquin Provincial Park in Ontario and its location in North America

*the true Algonquins by ancient writers, lived at Lake Nipissing.'*

Today, over a century later, a new term, *Algonquian* (or sometimes *Algonkian*), is used to designate the large group of related nations the commissioners cited in their report. The original French term, *Algonquin* (sometimes anglicized to *Algonkin*) is reserved for the particular Algonquian-speaking peoples who lived throughout the Ottawa River watershed that occupies what is now western Quebec and eastern Ontario, including most of today's Algonquin Park.

Of course, native occupation of the Algonquin Highlands began long before the arrival of Europeans. Seven thousand years ago, nomadic Paleo-indians arrived, probably exploiting the fisheries in fall and occupying traditional, family-held hunting territories each winter. These early people gave way to others and a succession of different cultures followed, culminating with the Algonquins whom the French found in the early 1600s. Soon afterwards, however, the Algonquins, the Nipissings and other neighbouring Algonquian-speaking nations were attacked and scattered by Iroquoian-speaking nations from the south. The invaders held sway until the early 1700s when they were forced to retreat back to their ancestral lands south of Lake Ontario by the Ojibwa, an Algonquian-speaking nation from north of Lake Huron. Descendants of the Ojibwa still live in southern Ontario today. As for the Nipissings and Algonquins, after being routed by the Iroquois, they lived in a succession of French missions in Quebec, finally ending up at Oka, just outside Montreal, and a few of their intermarried descendants eventually returned to scattered farms and villages in the Ontario part of the Ottawa Valley in the 1800s, including to what eventually became the Golden Lake reserve, southeast of what is now Algonquin Park.

By this time, other people were present in the Ottawa Valley as well. Most importantly, pioneer loggers had followed in the footsteps of explorers and fur traders. Fuelled by the British demand for pine timber, the loggers pushed into the far reaches of the Ottawa watershed, reaching what is now eastern Algonquin Park in the 1830s. Hard-working loggers spent winter after winter in the isolated highlands. They felled and squared countless trunks of pine and floated them in huge log rafts down the Ottawa River to the St. Lawrence River, and on to Quebec City. There the lumber was shipped across the Atlantic to its final destination in Britain. A few decades later the production of sawn lumber, in response to demand from the United States, took over. By the end of the nineteenth century most of the ancient forests had been devastated, either by logging or by fires.

In 1879 the first idea of protecting these forests arose, and in 1893 the Ontario Legislature passed the legislation establishing Algonquin National Park. People, like Robert W. Phipps, Alexander Kirkwood and James Dickson, who had surveyed the area of the park, and George MacCallum, are Algonquin Park's founding fathers. Algonquin Park was to be *'a public park and forest reservation, fish and game preserve, health resort and pleasure ground for the benefit, advantage and enjoyment of the people of the Province.'* In 1913 the name of Algonquin National Park was changed to Algonquin Provincial

INTRODUCTION

**Above:** *Moose, a native animal, and the largest of the fifty-three mammal species recorded in the park*
**Left:** *Purple Loosestrife, an invader, is one of the 283 non-native plant species which have established themselves in the Algonquin environment.*

Park, since the park had always been under the jurisdiction of Ontario. Timber remained the main source of revenue and logging was regulated only later, when it conflicted with recreational activities. Today, under a forest management plan, logging continues in much of the park. Compared to the pre-timber era, Algonquin (especially the west side) looks different, lacking the White Pine giants that were growing scattered in the hardwoods in thousands. However, according to research studies, the logging in the past as well as today has only passing and minor effects on the forest structure and composition and on the wildlife.

The era of tourism began with the arrival of the railway, built mainly for the purpose of timber transportation, followed later by the construction of Highway 60. The numbers of park visitors have increased since those humble beginnings to tens of thousands per year. The park has always attracted the attention of different human interests. Primarily it was logging, which under controlled management is still carried out today.

## ALGONQUIN PARK: A PORTRAIT

Subsequently, recreation and tourism have played an important role. It has also been the object of many research studies and conservation projects. Algonquin Provincial Park is one of the most researched natural areas in the world.

Algonquin's richness of wildlife is unique and diverse, due mainly to the park's location. The park lies at the zone where the southern deciduous forest meets the northern boreal forest, where the southern species of plants and animals meet their northern neighbours. Specifically, the 760,000 hectares host over a thousand species of vascular plants, about the same number of fungi, an estimated seven thousand species

**Below:** *Early park visitors*
**Right:** *Honking Canada Geese passing overhead. Today, the Canada Goose is one of about 140 breeding species of birds in Algonquin, but in early park days its sightings were limited to migrating flocks in spring and autumn.*

# INTRODUCTION

of insects, about 140 species of breeding birds (not counting another hundred that are rare or appear only in migration), fifty-three different mammal species, fifty-four fish species, seventeen species of amphibians, and fourteen species of reptiles. The diverse habitats of the park support a wide variety of species and that is what many people appreciate.

# SPRING *Awakening of Life*

Hardwood Lookout Trail in Algonquin Park, *painting by Jan Rinik*

# ALGONQUIN PARK: A PORTRAIT

Of the four seasons familiar to those of us who live in the temperate regions of the world, spring is the most awaited, especially after the long northern winter. The winters in Algonquin are long. From November until the beginning of April snow, frost and chilly winds rule the park. But then with the first warm sunny days of spring, great changes take place and life in Algonquin awakens once again.

The first spring changes go almost unseen. The Sugar Maple's sap starts to run when snow is still around and temperatures may still drop down below freezing. The sap brings back nutrients stored during winter in the roots and helps develop the leaves and flowers. Many tree flowers are quite inconspicuous, green and brown, but their role is the same as that of the most attractive and colourful plant blooms: the securing of a new generation. Spruce, pine and other evergreens also produce clouds of small pollen grains.

Among the most visible spring changes are those in the Algonquin hardwood forest, as this is the season of wildflowers. During this relatively short period, before the trees leaf out, carpets of lovely flowers appear to cover the forest floor. Red and Painted Trilliums, Spring Beauty, Trout Lilies and many others bring vivid colours back to Algonquin. The woods are filled with the drumming of woodpeckers and Ruffed Grouse, and the trills of warblers, which have just arrived back from the south.

The life in lakes, rivers and ponds awakens as well. After the ice cover is gone, Common Loons return from their wintering areas along the Atlantic coast. Their loud, haunting calls proclaim the definite victory of spring over winter. Beavers

# SPRING – *Awakening of Life*

*Sugar Maples in spring, as seen from the Visitor Centre.*

*Yellow-rumped Warbler (Setophaga coronata), a male in breeding plumage.*

repair their lodges and dams and forage for fresh food, as do the other four-legged park residents. The most frequently seen is the moose, which survives winter eating twigs and tree bark, and once again has easy access to the ponds where it can feed on juicy aquatic plants. The temperature of the water in lakes gradually increases and streams and rivers, powered by melted snow, sweep away the last reminders of winter.

# ALGONQUIN PARK: A PORTRAIT

*Spruce and pine needles are the only winter diet of Spruce Grouse, the specialized vegetarian of Algonquin's coniferous forests.*

The long and harsh winters in Algonquin permit only a few birds to stay in the area; the majority of species migrate south in the autumn. One of the permanent residents is the **Spruce Grouse**, *Falcipennis canadensis*, a bird indigenous to the boreal forests of North America. This species can survive long winters thanks to its ability to eat spruce and pine needles, which are available all year round.

The Spruce Grouse is also one bird species that can often be spotted early in spring. The birds are tame and chances of seeing them are quite good. Either on the ground floor of the forest or in the lower branches of conifers, or moving between these locations with slow, noisy, almost vertical flights, these handsome birds will delight the observer. Typically they do not flee, since at this time of the year grouse are totally distracted by the need to mate and lay eggs. The preferred habitat in Algonquin is dense spruce growth often found at the edges of bogs.

SPRING – *Awakening of Life*

**Above:** *Spruce Grouse at Spruce Bog Trail*
**Right:** *A branch of Red Spruce with a cone*

*Yellow-bellied Sapsucker*

*Hairy Woodpecker*

In early spring, long before the arrival of many other bird species, the drumming and tapping of woodpeckers is one of the most familiar sounds in the forest. Woodpeckers have bills used like a carpenter's chisel. By rapidly hammering through the bark, they gain access to their food — mostly insect larvae. One species, the **Yellow-bellied Sapsucker**, *Sphyrapicus varius*, is even more specialized, and instead of eating insect larvae, it 'drills' small holes in the tree bark, one next to the other, to feed on the sap that flows from these holes. In the autumn, when trees cease sap production, the Yellow-bellied Sapsucker migrates south. These holes also become attractive feeding stations for the Ruby-throated Hummingbird, a tiny bird that arrives in May. It may seem surprisingly far north for a hummingbird to be found in the park, but this species thrives and breeds in Algonquin.

The medium-sized black and white **Hairy Woodpecker**, *Picoides villosus*, is a common, year-round park inhabitant. It frequents hardwoods, climbing tree trunks up to the branches and using its strong bill to dig out insects, bracketing itself on the trunk or branches with its stiff tail feathers. The adult males have a distinct red nape that is not found on females.

SPRING – *Awakening of Life*

The Algonquin woodlands, with their rich mixture of deciduous and coniferous trees, provide a good habitat for several other woodpecker species. Some of them stay in the park all year. Others, like the **Northern Flicker**, *Colaptes auratus*, depart for the winter. The Northern Flicker often forages on the ground for ants, picking them up with its long and sticky tongue.

Northern Flicker

# ALGONQUIN PARK: A PORTRAIT

*Hardwood Lookout Trail*

After a long winter, April is a month of change. It begins with cold days and melting snow, but as the days get warmer and longer, and frosty nights shorter, life slowly awakens. The month of May brings even bigger changes. Bare trees, which lost their colourful leaves the previous autumn, are getting all dressed up again. First the buds and inconspicuous green flowers of maples and then vivid green leaves spring out on twigs, and soon the whole forest is full of flourishing life. Many bird species are arriving back from the south, making the forests come alive with their lovely songs.

SPRING – *Awakening of Life*

In the hardwood forests of Algonquin Park, before the canopy of dense leaves is complete and blocks most of the sunshine from reaching the forest floor, spring wildflowers bloom. They may be small or subtle but their beauty is awesome. One of them even bears a common English name that says it all, Spring Beauty.

Everybody in Ontario is familiar with the **White Trillium**, *Trillium grandiflorum*, as it is the symbol of the province and a widespread spring wildflower in the hardwood forests of southern Ontario. Nevertheless, it is rare in Algonquin, perhaps because of the enormous numbers of White-tailed Deer that virtually destroyed ground flora in the park in the past, including the White Trillium and other wildflowers (see pages 138–139).

There are two other Trillium species that are quite common in Algonquin Park. Both the **Red Trillium**, *Trillium erectum*, and the **Painted Trillium**, *Trillium undulatum*, are flowers of the Algonquin hardwood forest floor. Their conspicuous blooms appear after the snow has melted and before the forest floor is shaded by the leaves of trees. The Painted Trillium is similar in appearance to the White Trillium except it has radiating red streaks.

*White Trillium*

*Red Trillium*

*Painted Trillium*

# ALGONQUIN PARK: A PORTRAIT

*Northern Spring Beauty*

*Woolly Blue Violet*

The subtle flowers of the **Northern Spring Beauty**, *Claytonia caroliniana*, are one of the earliest, appearing in moist deciduous woods. The beautiful white or pinkish petals with darker striped veins close at night and during rainy weather, opening again on the next sunny day. The flowers will last only a few days. The little seeds, covered by oil, attract ants that carry them to different parts of the woods, ensuring their distribution. The flowers bloom within a brief period after the snow disappears and before the trees get their leaves. The scientific name of the species honours John Clayton, an eighteenth-century American botanist.

Another group of flowers, small but well known for their colours, are violets. From about five hundred different species worldwide and about sixty in North America, some sixteen species occur in Algonquin. The **Woolly Blue Violet**, *Viola sororia*, is a regular early spring flower of moist forest swamps and meadows. This plant forms charming, ground-covering bunches with purple, blue or lavender flowers. As in any other violet, the flower has five unequal petals, the lowest one being a landing pad for bees and other insects. However, many of the showy violet flowers are not pollinated, and later in the season a second set of flowers appear, tiny and close to the ground. These are guaranteed to produce seeds because they are self-pollinated. Many violets, especially the European Sweet Violet, are popular for their attractive smell, and are used in perfumes. Most of the North American species have little to no smell.

SPRING – *Awakening of Life*

Trout Lily

The **Trout Lily**, *Erythronium americanum*, is a frost-resistant early spring flower of the maple forest. No sooner has the snow melted when the pointed and spotted leaves penetrate upwards through the winter-flattened maple leaves, heading towards the first warm rays of the sun. The mottled leaves are said to resemble the markings of Brook Trout. The yellow nodding flower heads grow from underground bulbs that reproduce themselves by creating new daughter bulbs. This is the reason why these plants seldom grow singly. Instead, the Trout Lily typically covers large patches of the forest floor, making it very pleasing for the human eye. They can be seen at Hardwood Lookout Trail, as well as in other west side hardwood forests.

# ALGONQUIN PARK: A PORTRAIT

**Left:** *Beaver dam at Amikeus Lake*
**Below:** *Beaver lodge. Dams and lodges, signs of the presence of beaver in Algonquin Park, are usually solid and massive.*

The **North American Beaver**, *Castor canadensis*, is a unique animal and one of the most attractive in Algonquin. It changes its environment to make it more suitable for its own survival. By building dams on a small stream, beavers can transform a forest or meadow to a pond full of water. The warm shallow water of the pond will soon host many more plants and wildlife. Therefore, it is not only the beaver that benefits from the changes, but a whole array of other creatures. The beaver is the most amazing animal in Algonquin and beaver ponds are one of the park's characteristic ecological features.

SPRING – *Awakening of Life*

**Above:** *Beavers favour the bark of aspens and willows.*
**Left:** *Although it is not easy to observe feeding beavers, they can occasionally be seen swimming.*

# ALGONQUIN PARK: A PORTRAIT

Beavers are very shy and usually are not easy to spot. They have a good sense of hearing and any sound will often make them quickly dive under water. Beavers are active all year round and, especially in the fall when they are storing food for winter, they really are 'busy as beavers'. They remain active under the pond's ice cover in the winter. For the rest of the year, they repair their dams and lodges using twigs, branches and mud. Beavers are well known for cutting down trees, the signs of which are visible all over the park. They feed on the bark of their favourite trees: aspens, willows, birch or other deciduous species. They have sharp teeth that never stop growing. To keep them at the appropriate length, they must chew and grind wood all the time.

Among many extraordinary features that make the beaver well adapted to its special lifestyle is its flat tail. When a beaver detects danger, it dives with a slap of its flat tail, a warning signal to other family members. Besides its use as a rudder while swimming (especially when dragging the load of branches in its mouth), the beaver's tail is also important as a fat storage reserve in winter and body-heat-releasing device in hot summer temperatures. Though a graceful swimmer, the beaver is clumsy on land, where it can become easy prey for wolves. Inside its lodge, where the mother rears her kits, the beaver is safe.

The beaver was very important in Canada's early history. It was beaver pelts and the great demand for fur in Europe that prompted fur traders to explore the interior of the North American continent. Many traders and native trappers were involved, and by the beginning of the twentieth century beavers had disappeared from many places. The area of today's Algonquin Park was one of the places where beavers were trapped, and at the time of park's establishment in 1893 their population was very low. However, they soon recovered from over-trapping to the point that in the 1910s many of them were live-trapped and used to restock other areas, especially in the US, where they had been wiped out. Beavers continue to be widespread in Algonquin, although considerably less abundant than in the park's early days when logging and forest fires created ideal conditions for their preferred food, the bark of aspen trees.

**Above:** *The great demand for beaver pelts employed many fur traders and native trappers. After the beaver was trapped most of the fur preparation work was done by women, including stretching the fur on wooden frames.*

**Right:** *From the seventeenth to the nineteenth century, beaver hats were stylish headpieces in Europe and in the New World. They were favoured by military as well as by civilians.*

SPRING – *Awakening of Life*

# ALGONQUIN PARK: A PORTRAIT

There is no more distinctive sound in Algonquin Park than the call of the loon. For many people it is a symbol of the wilderness and an invitation to embrace nature — although, particularly with the yodel call given by males, it can be a deadly threat against rivals. Many of the loon's calls are loud and resonant and can even resemble the howl of a wolf. Newcomers to Algonquin sometimes fail to realize that the calls are made by a bird!

With its body length of ninety centimetres and a weight of up to 4.4 kg, the **Common Loon**, *Gavia immer*, is similar in weight and body length to a Bald Eagle. It has a robust, sharp bill and excellent swimming and diving abilities. It catches fish by swift underwater pursuit using its large webbed feet, holding its wings tightly close to the body. The loon can emerge close to the place where it submerged but also unpredictably far away, disappearing from the sight of a patiently waiting observer.

**Above:** *Common Loon on Opeongo Lake*

Open lakes are the loon's habitat, and its breeding sites are often small islands. The nest is built near the water's edge and usually two chicks are raised. Soon after hatching, the chicks abandon their nest and swim with their parents. Sometimes they can be seen resting and 'sailing' on a parent's back. Mated pairs are usually faithful to each other and stay together for as long as both birds are alive.

The loon's wild laughter can be heard in Algonquin from the end of April until the end of November. When the lakes freeze loons migrate south, spending winter on the Atlantic coast. In the early spring, soon after the ice melts, they return to the same lake and after a short courtship will start breeding again.

Common Loons are a familiar sight on many Algonquin lakes from early spring to late autumn.

Early spring in Algonquin Park, *painting by Jan Rinik*

# ALGONQUIN PARK: A PORTRAIT

*Roadside ditches along Highway 60 are attractive places for moose in spring because they are rich in sodium from winter salting operations on the highway. Moose viewing can be excellent in such areas, especially early or late in the day.*

Nothing makes wild places more unforgettable than the sighting of a mighty moose. That memory often remains deep in the mind of the happy observer for years afterwards. Truly, Algonquin Park is a special place with one of the highest concentrations in Canada.

With a clumsy and chunky body, disproportionately long head, ears and legs, it looks strange to an observer, but if you watch the **Moose**, *Alces alces*, in its natural environment you will soon be convinced how all of its adaptations serve a purpose. The moose is actually built for the environment where it lives. Very tall with skinny legs, its long, massive head allows it to feed on roots and juicy aquatic plants in boggy ponds in summer, as well as on twigs and small trees three to four meters high in the winter. Moose live and feed in ponds and bogs surrounded by thickets of alder, aspen, birch and spruce. These thickets are often so dense and impenetrable that it is hard to imagine how moose get through them. But the moose, with its five-hundred-kilogram body and extremely long legs, does it with ease and grace and it seems that no barrier is thick enough to stop it. The long legs are also extremely useful in moving through deep snow. Moose are tall animals, about two metres or more in height.

**Above:** *Cow moose and her one-year-old calf*
**Left:** *Young moose*
**Right:** *Moose foot-print*

ALGONQUIN PARK: A PORTRAIT

When in prime condition, adult moose can often defend themselves successfully against predators. Calves and old individuals are vulnerable to wolves and bears, however, and all age classes are subject to serious torment from insects. Biting flies and mosquitoes in spring and early summer attack moose relentlessly. Hot summer days also cause severe discomfort for an animal that, in Algonquin Park, lives near the southern limit of its range. Autumn brings relief from summer heat, but many moose will get infested with winter ticks. They grow until spring when their blood-sucking activity is so irritating

# SPRING – *Awakening of Life*

**Below:** *An adult bull moose in Algonquin can weigh over five hundred kilograms. Each year, during the summer months, male moose grow antlers with a span of up to 1.6 metres and a weight of up to twenty-five kilograms. Moose living in more northern parts of Canada and Alaska are even larger.*

that moose start to rub and scratch themselves in an effort to get rid of the irritation. An animal infested by several tens of thousands of ticks can lose a lot of hair and potentially die if the weather turns wet and then very cold. Yet another significant factor that regulates the moose population in the park is a parasitic brainworm carried by White-tailed Deer that is harmless to them, but fatal when transmitted to moose. In the past, when deer numbers were high, moose numbers were consequently very low. In contrast, now that deer numbers are low, moose numbers are often high, sometimes reaching as many as four thousand animals.

Bull moose have palmate antlers that finish growing in early September, just before the rutting season. The bull (male) moose attempts to dominate rival males by presenting his antlers, and only rarely do fights occur with locked antlers and pushing contests. Later, around the New Year, the antlers fall off and the new ones start growing again in spring.

Moose, with their long legs, readily enter boggy ponds in order to feed on sodium-rich aquatic plants. These are important to their diet after a long winter of eating plants low in nutrients. In May, the female (cow) moose gives birth to one or two calves, often on islands or some other place safe from wolves and bears.

# ALGONQUIN PARK: A PORTRAIT

*The Red-winged Blackbird builds its nest in wetland vegetation, often close to the water surface. The young are fed insects, but for the rest of the year this species can feed on seeds as well.*

Although not as common as in the southern Ontario, the **Red-winged Blackbird**, *Agelaius phoeniceus*, is a characteristic bird of park wetlands. From early spring the male's cheerful, gurgling song is echoed in the reeds followed by the energetic dance from perch to perch, scaring off rivals and showing off his colourful virtue to females. These birds have beautiful plumage, the whole body being glossy black with two distinct bright red and yellow patches on their wings. However, the females (above) are much more modest in their dress, being inconspicuous to potential predators and blending well with their surroundings.

# ALGONQUIN PARK: A PORTRAIT

Many people stop to admire carpets of flowers in bloom, after spotting a bird, butterfly or an animal, but few will pay attention to the forest itself. The forest is the home of large towering trees and small plants, as well as tiny, unseen creatures, surviving on dead leaves or in rotting logs. It is also the home to well-known species such as squirrels, deer and wolves, as well as lesser-known bats and mice. All the diverse forms of life have their own roles and rich life histories. None of them could exist alone. Instead, they are all tied to each other in a complicated symbiotic web of life, which scientists are just beginning to understand.

Most of the forests on the park's west side consist of broad-leaved hardwood trees. In spring, rising sap transports minerals and sugar from the roots and prepares the trees for sugar production. The Sugar Maple, along with Red Maple, American Beech, Yellow Birch, Black Cherry and others, are green factories employing countless work units, the leaves. Leaves produce sugar, the source of energy on which virtually all other living things depend. First in the food chain are insects and other leaf and plant-eaters. Then other small creatures followed by large predators that prey on them. Thus, all these living creatures depend on the primary food source produced by trees and other green plants. Additionally, we all benefit from the oxygen that is released by the leaves.

*Algonquin Provincial Park is close to the northern limit for many hardwood tree species. The deciduous forests of eastern North America originally stretched from the park area all the way down to Florida but today most of that forest is gone forever, having given way to urban areas and farms.*

*The Sugar Maple is the dominant tree of the hardwood forest. Its sap was used by native people, who boiled it down and produced syrup. The first European settlers soon learned from the Native Americans how to produce syrup and sugar from maple trees. The late winter and early spring was the 'sugaring season,' involving collecting sap, drop by drop, from tapped Sugar Maple tree trunks, and then boiling it in large metal vessels until the desired consistency of maple syrup was obtained. Involving the whole family, the 'sugaring off' quickly became an important part of the lives of the European settlers, and it still remains an important Canadian industry.*

SPRING – *Awakening of Life*

**Above:** *Sugar Maple forest in Algonquin Park*
**Below:** *Sap dripping from a tapped tree*

When the first Europeans arrived, they found a wealth of large trees as a source of lumber for their daily life requirements. After the White Pine old-growth stands were cut down in the early 1900s, the loggers fixed their attention on the hardwood trees in what is now the west side of the park. As many as ten sawmills were built within the boundaries of the park and six railroads that transported Algonquin timber. The Algonquin Logging Museum at the east gate of the Highway 60 offers insight into park logging, from its early days in the 1800s right up to the present.

# ALGONQUIN PARK: A PORTRAIT

*Chestnut-sided Warbler*

*Nashville Warbler*

*Black-throated Blue Warbler (male)*

Warblers, though small, are beautiful birds that enliven the green spring woodlands of Algonquin with their bright colours and songs. Most of them return during the first warm days of May, and by the end of September they are migrating southwards again. This distinguished group of birds includes highly active inhabitants of the broad-leaf and coniferous forests in the North American continent. However, warbler species are not always easy to tell apart. Males and females usually have different plumage, and their young, after leaving the nest, are also different. Later in the summer, most of them change again, moulting into less striking fall plumage.

Warblers forage in trees, often high in the canopy, for insects. They are very musical in spring, when establishing and maintaining their breeding territories, but in the hot summer days many of them are silent and will go unnoticed except by the keenest of observers.

Male and female **Black-throated Blue Warblers**, *Setophaga caerulescens*, are very different and they look like they could be two different species of birds. While he has a distinct black throat and greyish-blue upper parts, seen in the photograph, she is more or less uniformly brownish-olive.

*Black-throated Blue Warbler (female)*

SPRING – *Awakening of Life*

The **Yellow-rumped Warbler**, *Setophaga coronata*, nests over a large portion of North America and the eastern race was formerly known as the Myrtle Warbler. It has a white throat and a few other characteristics that make it different from the yellow-throated birds of the same species found in western North America.

A male **Chestnut-sided Warbler**, *Setophaga pensylvanica*, as its name indicates, has two broad chestnut stripes along each of his flanks, which are less visible in females.

The grey-headed bird with a yellow throat and belly is a **Nashville Warbler**, *Oreothlypis ruficapilla*. These warblers build their nests on the ground in damp forests.

Chestnut-sided Warbler

Yellow-rumped Warbler

Northern forests are home to several raptor species and one of the largest in the park is the **Red-tailed Hawk**, *Buteo jamaicensis*, a powerful bird with a distinctive red or chestnut-coloured tail. A pair of Red-tailed Hawks builds a large nest high in the trees or on rocky canyon walls and raises one to three chicks in a season.

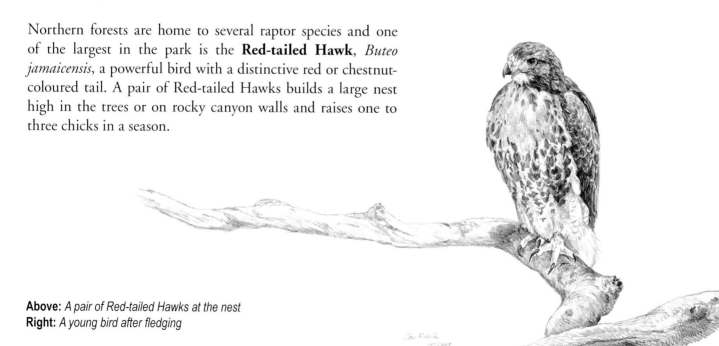

**Above:** *A pair of Red-tailed Hawks at the nest*
**Right:** *A young bird after fledging*

SPRING – *Awakening of Life*

Another, smaller but more numerous hawk in Algonquin Park, is the **Broad-winged Hawk**, *Buteo platypterus*. It is a typical forest raptor that seeks its prey by perching in a tree and suddenly dropping down to seize it. Snakes and small mammals are its favourite foods. The Broad-winged Hawk is a migratory bird, leaving breeding areas in September. Their migration is spectacular, with thousands passing along favoured cliffs or along the shores of the Great Lakes on their way south to Brazil.

**Above left:** *A flight silhouette of the Broad-winged Hawk*
**Above right:** *An adult Broad-winged Hawk perching at the top of a Black Spruce in Algonquin Park*

# ALGONQUIN PARK: A PORTRAIT

Osprey

Sharp-shinned Hawk

The combination of Algonquin's plentiful lakes and forest makes Algonquin Park an excellent home for the **Osprey**, *Pandion haliaetus*. The Osprey is a beautiful raptor with a white underbelly. Its diet depends almost exclusively on fish, and thus it must have access to open water. April, when the ice melts, is just the time when Ospreys return to Algonquin lakes. Hunting Ospreys usually hover high in the sky and after spotting a fish plunge down with outstretched legs. Their feet and talons are specially adapted to carry slippery and often still-living fish.

Other raptors that breed in Algonquin include the **Merlin**, *Falco columbarius*, a falcon species which catches small birds in the air.

The **Sharp-shinned Hawk**, *Accipiter striatus*, is the smallest hawk found in Algonquin but it is a very fierce hunter of small songbirds. It will pursue them right through thickets and forest, as it is a very agile flier. In the autumn the Sharp-shinned Hawk follows the migration of its prey southwards.

Young Ospreys at the nest

SPRING – *Awakening of Life*

*Merlin waiting on a look-out post for a small passerine bird to appear from the forest at the shore of Costello Creek*

The **Ruffed Grouse**, *Bonasa umbellus*, is a quite large bird but thanks to its nondescript, brown-coloured plumage matching well with old foliage on the forest floor, it often escapes our notice. Certainly in spring the bird is more often heard than seen. Its dull and low drumming, slow at first then accelerating more and more into a final frenzy, creates an eerie and mysterious atmosphere in the woods. The drumming is done by males trying to attract females. A displaying bird usually stands on a log and violently beats his wings in the air in order to produce his impressive performance.

**Above:** *Ruffed Grouse at Centennial Ridges Trail*
**Below:** *The detail of its tail feather*

# SPRING – *Awakening of Life*

The Ruffed Grouse is a vegetarian. In the winter it survives on aspen buds and later in summer feeds on berries and other fruit. The widespread **Bunchberry**, *Cornus canadensis,* is one of the favourite summer foods of Ruffed Grouse. Bunchberry often carpets the coniferous forest floor and its white flowers turn into ripe red berries in July.

*Bunchberry*

# ALGONQUIN PARK: A PORTRAIT

The **Pink Ladyslipper**, *Cypripedium acaule,* is a remarkable orchid. It has four twisted, purplish-brown petals and a pouch-shaped petal resembling a slipper, hence the common names Ladyslipper and Moccasin Flower. It grows in coniferous and mixed wood forests, in damp and mossy soil. Its life cycle spans ten years from germination to flowering.

The **Canada Mayflower** or Wild Lily-of-the-Valley, *Maianthemum canadense*, has zigzagged stems with small white flowers that turn into berries in midsummer. Ruffed Grouse and chipmunks often eat these berries.

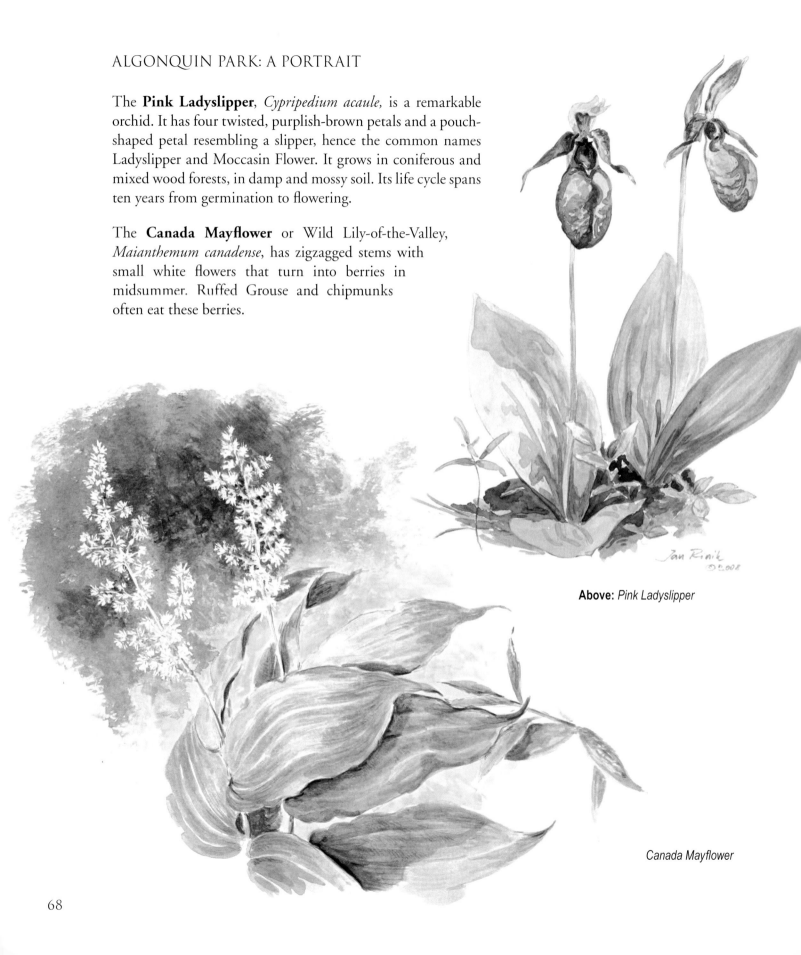

**Above:** *Pink Ladyslipper*

Canada Mayflower

SPRING – *Awakening of Life*

The cool moist shade of an Algonquin Park forest floor creates excellent conditions for the **Bunchberry**, *Cornus canadensis*, a diminutive member of the dogwood family. The low, ground-spreading plant often covers large areas in the forest. The flower heads have conspicuous white structures called bracts, with small greenish flowers in broad clusters, in the centre. They bloom in June and July, and later in the season the bright red, round and berry-like fruits appear.

Berries form an important part of many animals' diets. Animals that favour berries range from the Deer Mouse, chipmunks, squirrels and many birds, up to larger animals including Raccoon, Red Fox, White-tailed Deer and Moose. The Black Bear is well known for eating berries, thereby gaining valuable fat stores before the long winter (see pages 154–155).

Bunchberry

# ALGONQUIN PARK: A PORTRAIT

Chipmunks are among the best-known little creatures in Algonquin Park. It seems that people never get tired of watching them, for they are funny and entertaining. For a little while they may sit motionless, just gnawing something, and then suddenly they will dash, lightning-fast, to somewhere else only to stop and gnaw again. They often show their presence to campers, coming close to their feet, with the hope of getting a small piece of lunch. If they are successful, they run away until, a little while later, they come back with the very same request.

**Left:** *Twig and acorns of the Bur Oak,* Quercus macrocarpa, *a rare species in Algonquin Park*
**Below:** *Acorns of the Red Oak,* Quercus rubra, *are very nutritious foods for chipmunks, as well as for other wildlife.*

SPRING – *Awakening of Life*

The **Eastern Chipmunk**, *Tamias striatus,* is a small ground squirrel with beautiful dark and white stripes on its back. As a hardwood forest resident, the chipmunk likes maple seeds along with fruit, grass, mushrooms and insects. Before winter comes, chipmunks get busy harvesting and storing their food supply. All day long they run down to their underground storage chambers with maple seeds and hazelnuts in their large cheek pouches. They sleep through the long winter, but wake from time to time to visit their well-stocked food pile.

Chipmunk at the log in a picnic area in Algonquin Park.

# ALGONQUIN PARK: A PORTRAIT

The **Raccoon**, *Procyon lotor*, is a night-active (nocturnal) mammal, though it can also be observed during the day and evening. Aside from the campgrounds, the raccoon's preferred habitat in Algonquin Park includes water edges and lake shores, usually with dense vegetation and large trees where they will rest during the day. Often hibernating in hollow trees, raccoons overcome the low food availability during the snowy winter by sleeping through it. The early spring is a hard time for them, since food is still scarce and they are very hungry. The raccoon is an omnivorous animal, eating everything available from eggs, birds, snails, mice, fish and crayfish to berries and fruit. Raccoons readily enter water and can swim, or climb up high trees, in order to get food.

The mother raccoon usually gives birth to her young in May; they will soon follow her on foraging trips. A mother with her five or six small, ring-tailed youngsters crossing the road is an appealing picture. In Algonquin Park they are likely to be seen from early April until late autumn.

Raccoon resting on a maple tree,
*painting by Jan Rinik*

# SUMMER *Days of Bounty*

Rutter Lake, *painting by Jan Rinik*

# ALGONQUIN PARK: A PORTRAIT

Algonquin summers are warm and pleasant, and tourists from all over the world appreciate them. Most park visitors schedule their vacations for that time of the year. The excellent canoe routes on lakes and rivers promise the ultimate experience of the wilderness, as do backpacking treks and camping in the woods.

Summer is the season of babies and the young. Birds are silent, no longer making their presence known by a chorus of songs. The new generation of birds have enough time to prepare for either the long migration to the south, or winter survival under the severe Algonquin conditions. Her previous year's calf may still accompany the mother moose, but as soon as she gives birth to the new one, the elder calf will be forced to leave her. Wolves and bears also rear their young, as well as chipmunks, squirrels and all other mammals.

The summer vegetation is a deep green, lacking the vivid colours of the spring. Tree leaves have some four months for the process of photosynthesis before they cease it again. Summer wildflower species are stronger and taller than the early spring ones, prepared to withstand the heat and drought. Some years are especially dry, and sudden lightning can bring fires. Some fires are devastating, burning away large forest patches. Nowadays, fires in Algonquin seldom occur, and when they do, most of them are quickly put out. However, fire has an important place in the ecology of the forest. Certain species of pines have cones that open up to release their seeds only in response to very high temperatures such as those that forest fires can produce.

The Algonquin summer is a time of insects such as leaf-eating caterpillars. They provide a welcome and abundant source of food for many species of birds. The park is also an

# SUMMER – *Days of Bounty*

ideal environment for pesky biting insects such as black flies and mosquitoes, which develop from larvae in water. Their presence, especially in early summer days, can sometimes be really difficult to manage. Thus, summer can bring many unpredictable changes to the park, each of which can have significant impacts on the forests and their inhabitants.

# ALGONQUIN PARK: A PORTRAIT

*Twelve-spotted Skimmer*

*Dragonhunter*

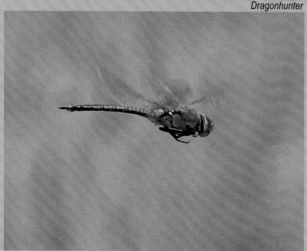
*Common Green Darner*

Of the many small creatures found in Algonquin there is a large group of insects that especially like warm summer days. Among them, park visitors in June will experience the first, but not very delightful, black flies and mosquitoes. Both types of insects can be extremely annoying, biting people and feeding on their blood. Black flies start their bloody business in May. Sometimes, they seem ever-present as they attack tourists in huge hordes as soon as one steps out of a car. Black and dark clothes are said to attract them, and their activity can last all day long. Mosquitoes come a little later in summer, and their highest activity is in the morning and in the evening. Both mosquitoes and black flies need blood to develop their eggs so only females attack humans and animals.

Other forms of insects more pleasant to see in the park are the dragonflies and damselflies. These are colourful creatures, glittering as they fly low over the still waters of ponds and rivers. They lay their eggs in the water or on aquatic vegetation, where their nymphs develop and feed. The larvae that hatch out, known as nymphs, are fierce hunters, sometimes even catching and eating small fish. When they mature, they come out of the water and transform into the adult stage. As adults they are aerial hunters, feeding on smaller insects.

*Chalk-fronted Corporal,*
*Ladona julia*

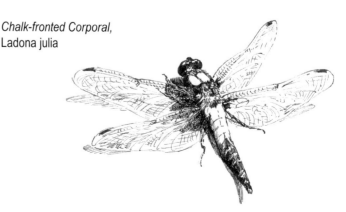

SUMMER – *Days of Bounty*

Mew Lake

Bullhead Lily,
Nuphar variegata

79

# ALGONQUIN PARK: A PORTRAIT

# SUMMER – *Days of Bounty*

**Left:** *Painted Turtles have characteristic yellow, orange and red markings. Thanks to their shells they are safe from most predators.*

Bogs and ponds, with their still and shallow waters, are the home of many aquatic plants. The **Bullhead Lily** or Yellow Pond Lily, *Nuphar variegata*, and the **White Water Lily**, *Nymphaea odorata*, are among the best known. Both species often create large green carpets of round leaves floating on the water surface. Their blooms are big and beautiful. The Bullhead Lily has yellow flowers, while the White Water Lily has snow-white blossoms. Many aquatic animals and fish find shade under the leaves, and frogs use lily pads to sit on and rest. The plants are nutritious and many different animals, from insects up to beaver and moose, like to eat them. Water lilies have their roots anchored in the bottom of the pond, and they often grow in quite deep waters.

Logs and stones are great places to bask for **Painted Turtles**, *Chrysemis picta*. From the early morning until the late afternoon they catch and absorb the warmth of the sun, which is so important for their metabolic processes. Unlike mammals and birds, turtles, frogs and snakes are cold-blooded, and dependent on external heat, since they can not produce it on their own.

*The Green Frog,* Rana clamitans, *is abundant and common in Algonquin Park.*

ALGONQUIN PARK: A PORTRAIT

*The garden irises have their wild relatives in the marshes and wet meadows of Algonquin Park. Blue Flag,* Iris versicolor, *is a tall aquatic plant with brightly coloured blossoms.*

Roadside flowers and those that grow in ponds or meadows reach their peak in midsummer.

SUMMER – *Days of Bounty*

**Right:** *The Pickerelweed,* Pontederia cordata, *is another aquatic plant that can be found in still waters of ponds and along lakeshores where it usually forms large clumps. It has blue flowers that bloom in July and August, and long stalks with lance-shaped leaves.*

**Left:** *Leaves of White Water Lily*

Of over a thousand plant species recorded in Algonquin, many grow in wetlands. Bogs, beaver ponds and still, shallow waters along the shores of lakes and rivers all have their own distinctive shrubs and wildflowers.

River Bank — Oxtongue River, *painting by Jan Rinik*

# ALGONQUIN PARK: A PORTRAIT

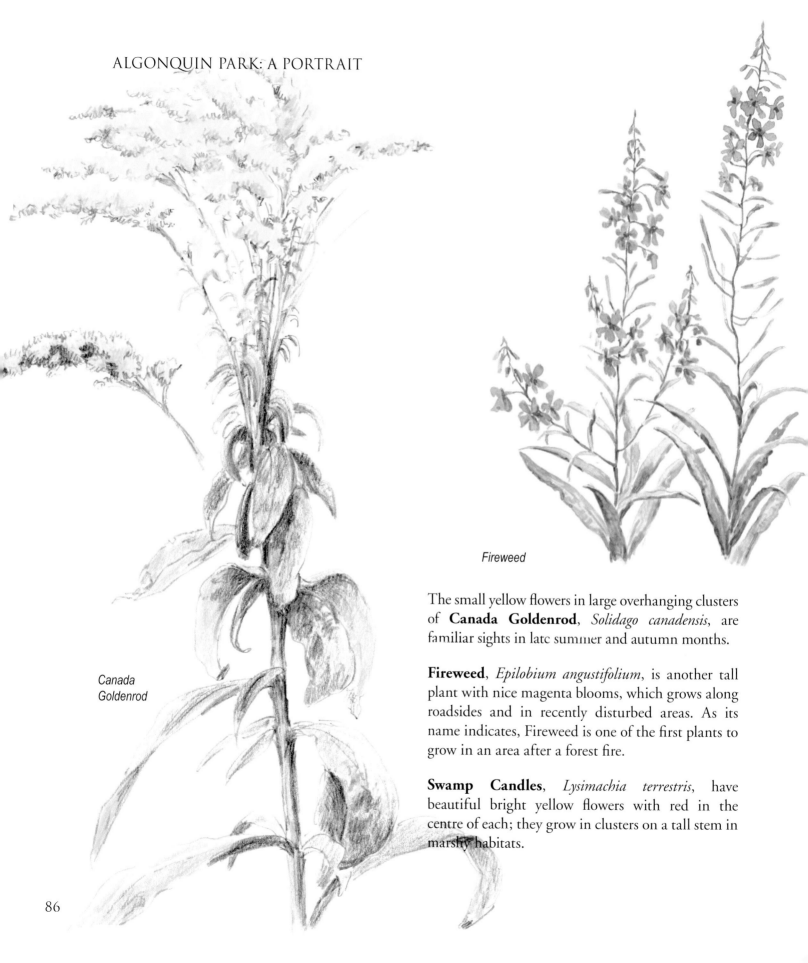

*Canada Goldenrod*

*Fireweed*

The small yellow flowers in large overhanging clusters of **Canada Goldenrod**, *Solidago canadensis*, are familiar sights in late summer and autumn months.

**Fireweed**, *Epilobium angustifolium*, is another tall plant with nice magenta blooms, which grows along roadsides and in recently disturbed areas. As its name indicates, Fireweed is one of the first plants to grow in an area after a forest fire.

**Swamp Candles**, *Lysimachia terrestris*, have beautiful bright yellow flowers with red in the centre of each; they grow in clusters on a tall stem in marshy habitats.

SUMMARY – *Days of Bounty*

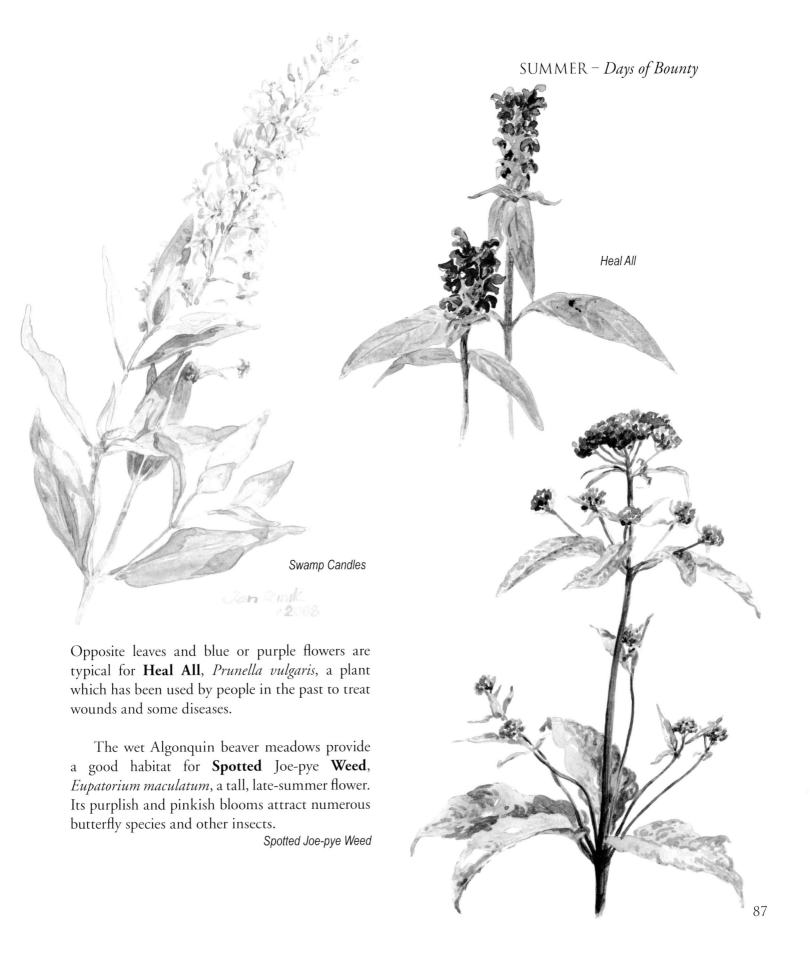

*Swamp Candles*

*Heal All*

*Spotted Joe-pye Weed*

Opposite leaves and blue or purple flowers are typical for **Heal All**, *Prunella vulgaris*, a plant which has been used by people in the past to treat wounds and some diseases.

The wet Algonquin beaver meadows provide a good habitat for **Spotted** Joe-pye **Weed**, *Eupatorium maculatum*, a tall, late-summer flower. Its purplish and pinkish blooms attract numerous butterfly species and other insects.

# ALGONQUIN PARK: A PORTRAIT

**Left:** White Admiral, Limenitis arthemis, *and other butterflies with their flimsy wings flittering from flower to flower and feeding on nectar are the adult forms upon which most species' descriptions are based. However, before they appear as adults, striking transformations take place as they go from egg to caterpillar to pupa. Different life stages have different demands and habitats but for all of them open and sunny sites are the best. Such locations are also the best places in Algonquin to watch butterflies.*

Insects are a very large group of animals, with seven thousand species estimated in Algonquin Park. Thanks to mosquitoes and black flies, insects often garner a bad reputation. However, within the class of insects there is an outstanding group that surely draws the admiration of all people. These are butterflies, the small flying jewels, pleasing humans all over the world with their bright colours and cheerful patterns. Algonquin's eighty species of butterfly may seem quite high given the park's location 'up in the north'. It is the variety of habitats and wildflowers found in the park that contributes to the diversity of butterfly species.

Many people recognize butterflies by their brilliant tones and appreciate their presence. However, many do not know that butterfly biology and life cycles are even more interesting and spectacular. Every single species goes through four stages of metamorphosis, from a single egg, then leaf-eating caterpillar, followed by pupa and finally the free-flying adult stage that people are most familiar with when they think of a butterfly. Adult butterflies are important flower pollinators, and many plants rely on them, attracting them with bright colours and sweet nectar.

*Question Mark,* Polygonia interrogationis, *is often feeding on rotting fruits, tree sap and even carrion.*

SUMMER – *Days of Bounty*

**Above:** *White Admiral*
**Right:** *Monarch on milkweed plant*

*Monarch larva*

One of the most spectacular phenomena in butterfly biology is the long-distance migration of the **Monarch**, *Danaus plexippus*. Adults spend winter and roosting in Mexico, and begin mating in early spring, still on their wintering grounds. Males will die soon and the females, while developing eggs in their bodies, will migrate northwards. On their journey, most of them will lay eggs and die. The second generation, emerging from eggs and accomplishing its transformation through larva, pupa and finally into adulthood, will reach Algonquin Park at the beginning of summer. Here, the adult Monarchs lay eggs on milkweed plants, where the resultant caterpillars will feed. Milkweed plants are poisonous to other animals, and the toxins retained in the Monarchs' bodies can save their lives by making them an unappetizing prey item for birds. When the autumn arrives, the generation of Monarchs already developed in Algonquin begin the long journey south back to Mexico.

ALGONQUIN PARK: A PORTRAIT

Common Grackle

American Crow

Black-coloured birds usually do not attract attention, but some black birds are well worth a second look. From a distance the **Common Grackle**, *Quiscalus quiscula*, looks evenly black in appearance. However, when examined close up and lit by the sun, its plumage shines with glossy and iridescent purple reflections. It has beautiful bright yellow eyes and a nice wedge-shaped, long tail. In spring, the Common Grackle is a vocal, but not very musical bird, producing brief, high-pitched squeaky noises. Its favoured habitats in Algonquin Park are ponds and wet woodlands.

Larger in size but similarly black in its appearance is the **American Crow**, *Corvus brachyrhynchos*. It is another ever-present and well-known bird. In Algonquin, readily emerging from trees, the crow patiently waits for food remnants in the picnic areas or elsewhere where people are present. Both the Common Grackle and American Crow will usually migrate southwards to avoid the harsh winters in the park. The much larger Common Raven (see page 96) is yet another black bird but is often seen circling high in the sky searching for prey rather than foraging in the style of the Crow.

**Right:** Spruce Forest along Oxtongue River, *painting by Jan Rinik*

# ALGONQUIN PARK: A PORTRAIT

*Barron River*

Many Algonquin Park visitors are familiar with the Highway 60 corridor and with the trails or canoe routes accessed from this highway. The surrounding rolling hills are known as the Algonquin Highlands and are the source of all the park's major rivers. The Highlands have elevations ranging from four to five hundred metres above sea level, and most of them are covered with hardwood forest in which the Sugar Maple is the dominant tree. Quite different is the landscape of the east side of the park. The forests are dominated by White, Red and Jack pine and the elevation is as low as 170 metres above sea level. The climate is different as well, with a warmer temperature and less rain. Most of the moisture falls in the Highlands portion of the park, as weather in Algonquin generally moves from the west to the east. The soil on the east side is sandy and many

**Above:** *The significant differences between the park's two sides are geological in origin. The melting glacier disgorged ground-up gravely material, the till, that forms the soil retaining moisture and supporting hardwood forest of the park's west side. The Barron River flows to the Ottawa River east of the park. Some eleven thousand years ago the Barron Canyon was overflowing with waters from a receding glacier. The two kilometre thick glacier melted and drained through the northern and eastern side of the park, forming the Fossmill Outlet. This huge river flowed for about five hundred years and deposited millions of tons of sand and gravel in low-lying parts of the east side of Algonquin Park.*

# SUMMER – *Days of Bounty*

**Right:** *The evergreen shrub, with prickly silver green needles growing on well-drained soils and rocky slopes on the park's east side is the* **Common Juniper**, *Juniperus communis. Juniper is a slow-growing conifer with berry-like cones and characteristic fruits; small and green initially turning a reddish colour when ripening and finishing a beautiful blue-black.*

trees growing in this area are drought-tolerant. Dominant species are White and Red Pines, Red Oaks and poplars. The east side covers about one third of the park and its approximate extent is marked on the map on pages 24–25.

*Twig of a Common Juniper*

*Pointed leaves of Red Oak, Quercus rubra, a characteristic tree species on Algonquin's east side*

# ALGONQUIN PARK: A PORTRAIT

*Barron Canyon*

The spectacular Barron Canyon, winding through the east side of the park (see map on pages 24–25), is a famous landmark and geological feature of Algonquin Park. It is located about a three hour drive eastwards from the Highway 60 part of Algonquin. The hard, crystalline rocks of the canyon are made of gneiss and granites. The concentration of iron in the canyon walls brings some reddish colour, in addition to greyish and off-white tones.

The cliffs of Barron Canyon offer specific breeding opportunities for insectivorous birds like the Barn Swallow and Eastern Phoebe. The **Barn Swallow**, *Hirundo rustica*, is familiar to many people for its habit of building nests on human structures — especially on walls of buildings and bridges — in much of Ontario and elsewhere. However, on the east side of the park,

SUMMER – *Days of Bounty*

**Above:** *Barn Swallow*
**Below:** *Eastern Phoebe*

in the Barron Canyon, it nested until the late 1990s the way the species did for thousands of years before taking advantage of human structures. Sadly, Barn Swallows and other birds that feed on flying insects have declined dramatically during the last fifteen years, and this swallow no longer breeds in the Canyon.

The **Eastern Phoebe**, *Sayornis phoebe*, a brownish-grey bird with a characteristic habit of wagging its tail, has breeding requirements similar to those of the Barn Swallow. Both species especially like cliffs where rock walls descend into the water. Outside the park, they tend to accept artificial structures provided by humans, and from this perspective, the natural breeding sites found in the Barron Canyon are unique and among the few known places where these species nest completely naturally.

ALGONQUIN PARK: A PORTRAIT

**Left:** *Common Raven feeding on a carcass*
**Below:** *Raven's silhouette in flight with a characteristic wedge-shaped tail*

Several other kinds of birds are ready to use cliffs when they are available. The black silhouette of a large bird with a wedge-shaped tail, often accompanied by harsh croaks, reveals the **Common Raven**, *Corvus corax*, a year-round resident in the park. The Raven is a big bird, larger than the similar and more familiar American Crow (see page 90). The steep walls of Barron Canyon are a favourite nesting area, but Ravens build on other cliffs and in big White Pines all over the park as well. During the harsh winter Ravens survive on carcasses and scraps left by wolves.

**Above and Below:** *Turkey Vulture soaring high in the sky*

Even larger and more conspicuous in its appearance is the **Turkey Vulture**, *Cathartes aura*. While soaring for hours, the vulture holds its long wings in a shallow V. Its head is bare and red, a feature often visible through binoculars. It comes as a surprise to some that vultures live as far north as Algonquin Park. Usually, these scavenging birds are more associated with the warmer and drier climates of the south, however, the Turkey Vulture has shifted northwards in the last few decades, possibly in response to global warming.

# SUMMER – *Days of Bounty*

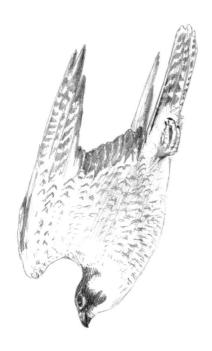

Nowadays the magnificent **Peregrine Falcon**, *Falco peregrinus*, is rarely seen in Algonquin Park, although it used to nest on some of the park's rocky cliffs. The decline of the Peregrine in the last century in North America and Europe is a sad story and a classic example of human impact on a fragile environment. The once-widespread bird ruled the skies as it plunged chasing its prey at speeds up to three hundred kilometres per hour but was brought to the brink of extinction by the introduction of the pesticide DDT to agriculture.

The Peregrine's principal prey, small birds, fed upon DDT-sprayed crops and poisonous organochlorines accumulated in their bodies. The falcon then ate these small birds, and the man-made chemicals caused the Peregrine to lay eggs with thin shells. Thus, the incubating female broke the eggs soon after they were laid and no young were produced. In an attempt to save this wonderful bird, governments and universities started breeding healthy birds in captivity in Europe and North America in the 1970s. After a few decades these projects resulted in several thousand young falcons being released back into the wild. Thanks also to concurrent restrictions on DDT, Peregrine Falcons are slowly recovering and re-occupying many of their former breeding sites — although, as of this writing, not yet in Algonquin Park.

*Peregrine Falcon stooping*

**Left:** Portrait of Peregrine Falcon, *painting by Jan Rinik*

After Rain in Algonquin Park, *painting by Jan Rinik*

# ALGONQUIN PARK: A PORTRAIT

*Great Blue Heron searching for fish at Costello Creek in Algonquin*

Algonquin's lakes, rivers and streams support several water birds including the large **Great Blue Heron**, *Ardea herodias*. The heron hunts patiently, motionlessly waiting and then suddenly seizing a fish or frog lightning fast with its bill. In Algonquin Park the Great Blue Herons can also be seen in trees where they roost and build their nests in small colonies.

SUMMER – *Days of Bounty*

*Belted Kingfisher on its perching post*

Another Algonquin angler is the **Belted Kingfisher**, *Megaceryle alcyon*. It is likely to be heard first before it is seen as it makes a loud, harsh rattling call as it flies from perch to perch, fast above the water. The Belted Kingfisher is a wary bird and usually will not let an observer approach closely. After spotting a fish the kingfisher dives headfirst into the water. If successful in its hunting foray, it will carry the fish in its bill back to one of its perching posts.

# ALGONQUIN PARK: A PORTRAIT

Mallard

American Black Duck

Ring-necked Duck

Algonquin Park lakes are deep, cold and surrounded by rocky shores. They are low in nutrients and do not support many waterfowl. The park's plentiful beaver ponds, however, are different. Their water is shallow, warm and highly productive, offering great quantities of food and sustaining many plants and wildlife, including ducks. The Mallard, the American Black Duck and the Ring-necked Duck are true Algonquin breeding waterfowl species that can be seen by park visitors from April until November.

The **Mallard**, *Anas platyrhynchos*, is well known from urban lakes and parks. The Mallard and the **American Black Duck**, *Anas rubripes*, are both found in Algonquin Park. Similar in size, the Black Duck, once common in Algonquin, is being replaced by the Mallard, spreading to the park area from the west. Both species are surface-feeding ducks with vegetation making most of their diet. The breeding plumages of adult male and female Mallards are much different from each other, with the male having a metallic green head, a chestnut breast and a yellow bill. The female is overall brown, resembling Black Ducks of both sexes.

The **Ring-necked Duck**, *Aythya collaris*, is a diving duck that searches for food on the bottom of shallow lakes, bogs and ponds. It has a distinctive angular head rather than rounded, with a high crown. Contrary to its name, the white ring around its bill is more visible than the barely discernible brown ring on the black neck of the male, from which the species got its name.

SUMMER – *Days of Bounty*

Canada Goose,
*painting by Jan Rinik*

The **Canada Goose**, *Branta canadensis*, is not as common in Algonquin as in southern locales, where it can number in the thousands. Only a few short decades ago most park sightings of geese were limited to migrating flocks. However, with the human introduction of Giant Canada Geese to southern Ontario, this race of Canada Goose has successfully adapted to city parks and rural landscapes and has subsequently spread north into Algonquin and become part of the local scene.

A Pair of Wood Ducks, *painting by Jan Rinik*

# SUMMER – *Days of Bounty*

**Above:** *Wood Duck perching in a tree branch*
**Below:** *A mother Wood Duck with her ducklings*

An extraordinarily beautiful bird, the **Wood Duck**, *Aix sponsa*, is a real marvel of Algonquin. Indeed, this bird is surely one of the most spectacular ducks in the world. The male's iridescent green, purple and blue plumage in combination with white is striking. The female's colours, as with many other wildfowl, are more muted and simple, brown and grey, providing camouflage from predators while they are incubating eggs.

The Wood Duck is a tree hole nester. The female will use Pileated Woodpecker holes, as well as other holes, and lay up to fifteen eggs. The nest hole can be very high in a tree, and the ducklings leave the nest soon after hatching by jumping down, following the call of their mother. Having the home high in the tree is a good protection against adverse weather and predators. The Wood Duck can be seen sitting on a tree branch more often than any other duck species. Being a shy bird, it delights the observer for only a little while, and then rushes off to safety.

SUMMER – *Days of Bounty*

Common Mergansers flying over the Oxtongue River, *painting by Jan Rinik*

109

# ALGONQUIN PARK: A PORTRAIT

Travel in Algonquin revolves around water. In the past, the network of rivers and lakes was the only way of travelling great distances for native peoples. Later, in the logging era, waterways were used to float the lumber down as far as Quebec City. Today, Algonquin rivers and lakes offer an excellent network for intrepid canoeists. Paddling a canoe and camping in the park's interior is an unforgettable and rewarding experience.

Six major rivers originate in the Algonquin Highlands. Meandering down from the park's west side, they join the Ottawa River in the east or Georgian Bay in the west. They are not big rivers, but afford good chances to see wildlife. On hot summer days and in the autumn their flow is slow, almost unnoticed, but they rapidly come alive in early spring at the time of snow melt. There are several rapids and waterfalls on the Oxtongue River, York River, Bonnechere River and others, but only the Petawawa is a true whitewater river. The Barron River, on the park's east side, flows through the magnificent Barron Canyon, one of the park's visual highlights (see page 94).

Algonquin rivers sustain diverse forms of life, from microscopic algae and bacteria up to the largest known fish in the park, the Muskellunge, which reaches up to one metre in length.

*In the past, the Algonquin people used the birch-bark canoe in their everyday lives. This canoe was light enough to be carried by one man around rapids, waterfalls or other barriers.*

Before sunset at Opeongo Lake,
*painting by Jan Rinik*

SUMMER – *Days of Bounty*

*Muskellunge*

The **Muskellunge**, *Esox masquinongy*, a member of the pike family, is a predatory fish. It is a native fish to Algonquin but only in the Petawawa River below Lake Travers on the east side and in the York River at the southernmost boundary of the park.

**Fallfish**, *Semotilus corporalis*, and **Creek Chub**, *Semotilus atromaculatus*, both of the minnow family, feed on other fishes and aquatic invertebrates. Fallfish are only found at the Petawawa watershed, not along Highway 60.

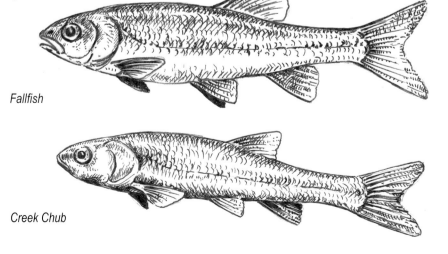

*Fallfish*

*Creek Chub*

SUMMER – *Days of Bounty*

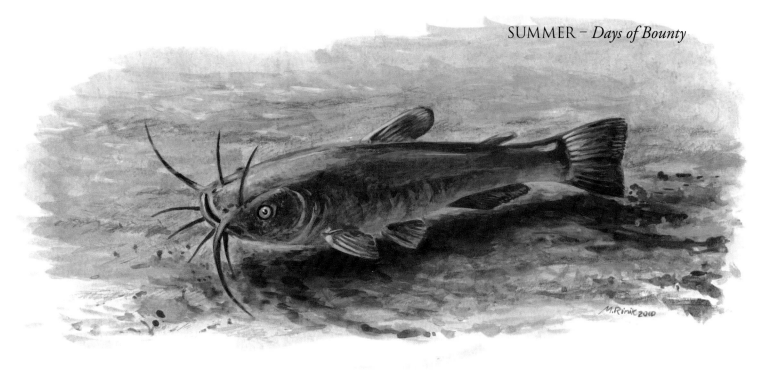

Brown Bullhead

**Below:** *The White Sucker,* Catostomus commersoni, *is a widespread fish of Algonquin streams, river and small lakes. It feeds on the mud-dwelling larvae of flies and other small organisms that it picks from the bottom with its distinctive sucking mouth.*

The **Brown Bullhead**, *Ameiurus nebulosus*, is a catfish with conspicuous barbels that help locate food (mostly small fish) in muddy waters at night. It has a massive head and stout body without scales. Its skin is extremely sensitive and able to detect chemicals from nearby food. Brown Bullheads, usually small, are less than twenty-five centimetres long, and live in most of park's rivers and lakes.

White Sucker

Algonquin lakes are beautiful and surrounded by woods. They're among the most attractive features of the park. Most of them are small or medium sized; yet a few, like Opeongo Lake, are quite large. Their waters are clear and, for most of the year, remain cold. In summer the upper level of water gets warmer while the lower and deeper parts stay cool, supporting many cold-water fish species.

One of the most distinctive features of Algonquin Park is its high and unique concentration of trout lakes. Formed when the last glacier melted northwards, these cold, oxygen-rich waters were ideal for trout, but unsuitable for warm-water species such as Smallmouth Bass, Northern Pike and Walleye. By the time the new Algonquin lakes and rivers had warmed up enough for such fish, there were no longer any barrier-free river corridors that would allow the warm-water fish to have access to the trout lakes in the Algonquin Highlands. Thus, for the last ten thousand years or so, lakes in the higher-altitude western part of the park have supported unique, trout-dominated ecosystems, largely free from competition from bass, pike and walleye.

**Above:** *In spite of its name, the Brook Trout,* Salvelinus fontinalis, *is found mostly in lakes in Algonquin Park. The key requirement is that there be suitable spawning beds — shallow gravelly areas with an upwelling of cold, well-oxygenated water to nourish the eggs as they develop over winter. Brook Trout have disappeared from many of the lakes into which Smallmouth Bass were introduced in the 1900s.*

**Above:** *Lake Trout*
**Below:** *Splake, the hybrid between the Lake Trout and the Brook Trout*

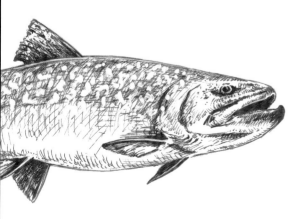

The **Lake Trout**, *Salvelinus namaycush,* one of the largest freshwater fishes in North America, reaches up to one metre long in Algonquin. It is a predatory fish at the top of a lake's food chain. It favours the deepest regions of the lakes, where it feeds on smaller fish like Cisco (Lake Herring) and Whitefish. It is a slow-growing species that matures and begins to spawn when it is about fifty centimetres long.

Brook Trout and Lake Trout do not hybridize naturally, but to improve fishing opportunities the hybrid between the Lake Trout and the Brook Trout, known as *Splake,* is raised at fish hatcheries and stocked in selected lakes, especially those with no native trout populations.

Northern Pike

The natural distribution of the **Northern Pike**, *Esox lucius*, was restricted to two lakes in the east side of the park. In the 1980s this fish appeared in the Opeongo River system, probably illegally introduced by people, and is spreading further upstream. The Northern Pike is a predatory fish and poses a serious threat to native fish communities, including the park's unique populations of Brook Trout and Lake Trout.

SUMMER – *Days of Bounty*

**Right:** *The Pumpkinseed,* Lepomis gibbosus, *or Sunfish, is an unmistakable, colourful fish widely distributed in the lakes and rivers of Algonquin Park. It is remarkable in that parental care and protection of newly hatched fish is performed by the male.*

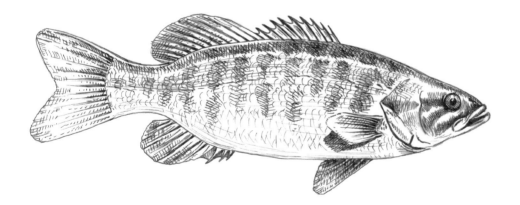

**Left:** *The Smallmouth Bass,* Micropterus dolomieui, *was originally restricted to the lakes and rivers south of Algonquin. It was introduced by an early superintendent to many of the lakes along Highway 60 and is now a common and popular sport fish. On the downside, it has displaced many populations of the native Brook Trout.*

**Right:** *The Yellow Perch,* Perca flavescens, *is one of the most widespread fish in Algonquin. It has vertical bars on its body, helping to camouflage itself in aquatic vegetation. The Yellow Perch grows rapidly, often out-competing Brook Trout and Splake by taking better and larger food items.*

# ALGONQUIN PARK: A PORTRAIT

# SUMMER – *Days of Bounty*

Elusive weasels, martens and fishers, small- to medium-sized woodland predators, are highly active and speedy. They forage along forest floors and tree branches for squirrels, chipmunks, mice and small birds or their eggs. Their good senses of smell and hearing allow them to flee before being seen by humans. Their dense fur coats have always been prized and they were trapped in large numbers over past centuries. This contributed to their serious decline, so much so that for years Algonquin Park (where trapping was not permitted) was one of the few places in Ontario where they could be found. Since then, Ontario's system of registered traplines has allowed martens and fishers to reclaim much of their former ranges. In search of food, the **American Marten**, *Martes americana*, often stops, stands on its hind legs and sniffs before moving off swiftly.

*Marten*

*Dawn at Amikeus Lake, painting by Jan Rinik*

# ALGONQUIN PARK: A PORTRAIT

Large coniferous forests, with their abundant supply of cones, play host to the **Red Squirrel**, *Tamiasciurus hudsonicus*, a striking little mammal. Red Squirrels possess marvellous climbing abilities and can reach the very top branches of a large pine or spruce in seconds to get their favourite food. White Pine seeds are among the favourites. Squirrels have a relatively small territory, where they know every single branch. If any intruder enters, the Red Squirrel will defend the territory vigorously. They are noisy, chattering animals, flicking their tails in an attempt to drive off bigger mammals and humans.

**Left:** *The rusty coloured Red Squirrel, rushing up and down the trees or sitting on its favourite branch is a characteristic of Algonquin forests. In late summer and in the autumn the Red Squirrels clip spruce and pine cones without eating the seeds first. They later come down the tree and hide the cones in underground middens to serve as their winter food supply.*

**Right:** *Black Spruce stands at West Rose Lake are a favoured Red Squirrel habitat.*

# ALGONQUIN PARK: A PORTRAIT

Various species of spruce are the dominant tree species in boreal forests across North America, Europe and Asia. In Algonquin Park, three spruce species are found that are quite similar in appearance, but each grow under their own specific conditions.

The popular Spruce Bog Boardwalk near the park's Visitor Centre traverses typical boggy stands of **Black Spruce**, *Picea mariana*. Black Spruce has adapted to extremely hostile, acidic environments with little oxygen or available nutrients. Slow growing, with a slender shape and usually not very tall, the Black Spruce may reach ten to fifteen metres in height. Its needles are quite short and the cones small and spherical.

Much taller and more impressive in appearance is the **White Spruce**, *Picea glauca*, especially when growing in open areas like the picnic areas in Algonquin. It also grows in forests scattered among other trees, both broad-leaf and coniferous, such as aspen, White Birch, Balsam Fir and pines. The sharp-pointed needles and shiny brown cones are longer than those of the Black Spruce. The cones produce little seeds that are an important food for birds and squirrels.

*All species of spruce have a pyramidal look with branches drooping down to withstand the heavy snow loads of northern winters. Although their developing seeds are protected in tough, woody cones, this is ineffective against seed predators like Red Squirrels that can rip the cones apart, or crossbills — specialized finches that can pry open the cone scales and extract the seeds. The spruces have evolved the counter-defence of producing very few seeds for several years in a row (keeping squirrel and crossbill numbers low) and then producing a massive cone crop that overwhelms the seed predators. White-winged Crossbills may gather in the thousands (see pages 190–191) to feed on the bonanza, but a few seeds will nevertheless escape and go on to produce new trees.*

*Black Spruce twig with cones*

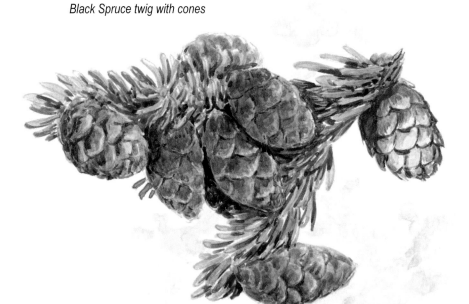

The least-noted spruce (and the only non-boreal spruce species in Algonquin) is the **Red Spruce**, *Picea rubens*. It usually grows scattered among Yellow Birch and Hemlock, having lower requirements for direct sunlight than other spruce species in the park. It is not a very common spruce in Ontario. A small stand of this spruce species can be seen on Red Spruce Side Trail next to the Hardwood Lookout Trail.

Black Spruce Trees and Mallards, *painting by Jan Rinik*

# ALGONQUIN PARK: A PORTRAIT

*Hemlock stand near Mizzy Lake Trail*

**Eastern Hemlock**, *Tsuga canadensis*, is a massive evergreen with stout branches that grows with other conifers along lake shores, especially on the west side of the park. Hemlock also forms distinct groves in upland hardwood forests. It likes cool places around lakes and on north facing slopes. Hemlock needles are flat, short and rounded at the tip, and the cones are very small — among the smallest of all local conifers. The trunks are massive and little else can be found growing under the trees, except for some saplings waiting for their turn when some of the old giants fall down to open the sky for a new generation.

**Right:** *Hemlock is an important tree for wildlife. In winter, its solid branches, with their thick foliage hold much snow, which is appreciated by deer. Under the hemlock the deer find a winter shelter, or deeryard, with less snow than under any other tree in the forest. White-tailed Deer will also eat the young Hemlock branches and needles. Breeding birds, such as Blackburnian Warbler, Setophaga fusca, also like Hemlock stands.*

SUMMER – *Days of Bounty*

**Eastern White Cedar**, *Thuja occidentalis*, prefers humid and swampy shores of lakes and ponds and the edges of bogs in Algonquin Park. In the nutrient-poor soils of Algonquin the cedar is dependent on places where nutrients are brought in by groundwater. Seedlings and the foliage provide good food for White-tailed Deer, once an abundant animal in the park, and the small seeds nourish birds such as the Red-breasted Nuthatch.

*Twig of Eastern White Cedar*

# ALGONQUIN PARK: A PORTRAIT

Indian Pipe

The translucent white plant called the **Indian Pipe**, *Monotropa uniflora*, is a specialist of the dark, sunless forest floor. With no leaves and no chlorophyll, the plant cannot produce its own food from sun and water through photosynthesis in the way most green plants do. Instead, it is a parasite. The clusters of plants employ little fungi and bacteria growing underground, around their roots, which draw nourishment and minerals from the decaying and dead leaves on the forest floor. Most green plants are visible from spring through to autumn, but Indian Pipe shows up only at the time of its reproduction, which is from July into August.

*Flowery Blewits,* Lepistra irina, *beside a rotting pine log. These are litter-decomposing mushrooms.*

SUMMARY – *Days of Bounty*

The **Eastern New**t, *Notophthalmus viridescens*, is one of seventeen species of amphibians in the park. After transforming from an aquatic larva, the conspicuous orange-red land stage (called an 'eft') lives several years on land before returning to water as a greenish aquatic adult.

**Right:** *American Robin*
**Below:** *Eastern Newt*

During the summer heat, the shaded forest floor provides food and shelter for many animals. Birds, like the widespread **American Robin**, *Turdus migratorius*, seek food there as well, especially earthworms. Robins are familiar to many people since they are commonly seen on lawns in our suburbs, but they are common in the wilds of Algonquin as well. They are migratory birds, as their scientific name reveals, leaving the park by the first frosty days of November.

Fungi are a very large group of organisms that have adapted to life in the dark forest. They lack green chlorophyll, the pigment responsible for photosynthesis, and thus do not produce sugar in the manner of green plants. However, they play a very important role in the life cycle of the forest. Together with bacteria and earthworms they are instruments of decomposition. Through complex biochemical reactions they turn organic litter, mainly dead leaves and trees, into rich humus. In this way they unwittingly prepare the forest environment for the next generation of life.

**Above:** *King Bolete,* Boletus edulis, *is a delicious mushroom. A similar look-alike, the Bitter Bolete, is too bitter to be edible.*

*Painted Suillus,* Suillus pictus, *a plum-pink mushroom growing in a pine litter layer, and leaves of Canada Mayflower*

# SUMMER – *Days of Bounty*

Over a thousand different species of fungi grow in Algonquin woods. Most of them are very small, usually invisible to the naked eye. The well-known mushrooms are those that send large fruiting bodies up into the air where we can see them. These fruiting bodies produce countless microscopic spores light enough to be carried far away by the wind. More mushrooms are seen in wet years than in dry ones. Mushrooms are part of the diets of many forest dwellers, from little insects to large bears.

**Above:** *The Birch Bolete,* Leccinum scabrum, *coming up through leaf-litter*

**Right:** *Yellow Swamp Russula,* Russula claroflava, *has a brightly coloured yellow cap and stout white stem.*

**Below:** *Golden Chanterelle,* Cantharellus cibarius

# ALGONQUIN PARK: A PORTRAIT

Eastern White Pine

The **Eastern White Pine**, *Pinus strobus*, significantly changed the history and economy of the Algonquin Park region. The abundant pine stands, with trees towering over forty meters, initiated pioneer logging beginning in the early 1800s. The king of commercial wood, the White Pine, was exceptionally versatile in its industrial uses and was made into everything from matches to furniture to ship masts. The demand for its lumber was great. The original giants on the west side that grew scattered in the hardwood forests were almost all cut down, leaving just smaller trees. One remaining old-growth stand can be seen at Big Crow Lake, but these more than three-hundred-year-old veterans are dying and falling down. They are not being replaced because the conditions for their germination and growth in the hardwood forest are extremely unusual.

The White Pine is a real monarch of the forest. Impressive individuals on the Big Pines Trail may be 220 years old. They have straight, massive, dark grey trunks; horizontal, widely-spaced branches and soft, almost feathery, bluish-green needles. The cones mature after two years on the branch. During the second autumn they shed their seeds and the cones

SUMMER – *Days of Bounty*

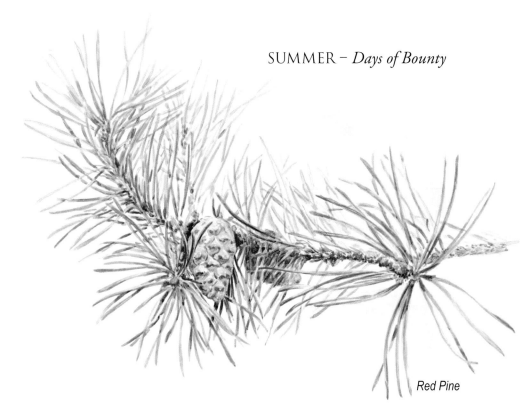

**Right:** *Along with the White Pine, the Red Pine,* Pinus resinosa, *is another potentially mammoth tree that is highly prized by the lumber industry in Algonquin Park. In Algonquin it has optimum conditions for its growth on the east side — the more sandy side of the park — where it thrives on sandy soils and gravely ridges. New generations of Red Pines usually get established thanks to forest fires, which burn away dense underbrush and expose mineral soil where pine seeds can germinate. Mature trees, on the other hand, are quite resistant to fires. The Red Pines never grow in hardwood forests of the west, unlike the White Pines that can grow well there if it can get started in the first place.*

Red Pine

fall to the ground. If a seed lands in an area of forest floor with good mineral soil, adequate moisture and sunshine, a new giant will start its life. Pines especially thrive on sandy soils on the east side of Algonquin, where hundred-year-old trees can be seen, often in almost pure stands.

*White Pine cone*

*Red Pine needles*

**Above:** *A Red Pine branch with green cones. Both White and Red Pines produce seeds on an irregular basis with good crops every three to seven years. The Red Pine has long needles bundled in twos instead of fives like the White Pine. The cones of the White Pine are up to twenty centimetres long.*

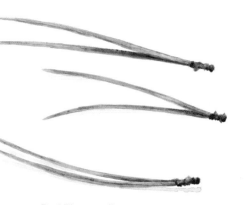

# ALGONQUIN PARK: A PORTRAIT

Jack Pine

If a natural, lightning-caused forest fire brings some benefit for the next generation of Red and White Pines, the **Jack Pine**, *Pinus banksiana*, is even more dependent on it. Indeed, this particular pine species is astonishingly well adapted to fire. The heat of the fire opens up the cones, some of which may have remained closed and retained their seeds for more than twenty-five years. Not only that, but fire also removes forest growth, releases nutrients and exposes mineral soil for the next generation of seedlings. Jack Pines are extremely rare on the west side of Algonquin, but are quite common on the east side.

*Jack Pine cone*

# SUMMER – *Days of Bounty*

*Forest fires were always a natural part of life in Algonquin, but, when loggers came, the frequency and extent of forest fires greatly increased. In the 1900s, in an attempt to combat fires through early detection, rangers built a network of towers from which towermen watched for tell-tale signs of smoke. Today, the towers are almost all gone and fires are usually detected by planes. An average of fifty fires start in the park yearly, mostly caused by carelessness of humans. About fifteen of them are started by lightning. Almost every fire in the park is now quickly put out.*

*Branch of Jack Pine with cones*

# ALGONQUIN PARK: A PORTRAIT

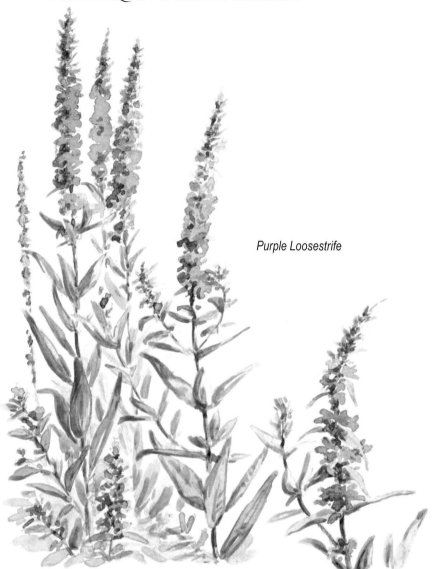

*Purple Loosestrife*

**Left:** *Purple Loosestrife,* Lythrum salicaria, *is an aggressive European wetland plant, introduced some two hundred years ago. In many places it pushes out native species, however, the plant has a restricted range in Algonquin and so far has proven to be little threat in the park.*

**Above:** *The Blueweed (or Viper's Bugloss),* Echium vulgare, *attracting a bumblebee*
**Below:** *The Blue Vetch,* Vicia cracca. *Both plants were originally found in Europe.*

Summer visitors in Algonquin Park can enjoy many beautiful wildflowers, even without hiking to the park's remote areas. Hundreds of species are in bloom just along Highway 60, at parking and camping sites, and at other places with good, sunny conditions. Most of these roadside species are not native to Algonquin. Some come from the prairies in the west, but most of them have a European origin. They were introduced by the first European settlers who established themselves in North America and eventually reached the park. They persist because they can exploit the conditions found along the highway better than most of our native flora.

*Blue Vetch*

# SUMMER – *Days of Bounty*

**Left:** *The showy Brown-eyed Susan,* Rudbeckia hirta, *native to the prairies, is well established along Highway 60.*

**Bottom Left:** *The aromatic green leaves of Yarrow,* Achillea millefolium, *have long been used in European medicine.*

**Bottom Right:** *The flowers of Yellow Toadflax,* Linaria vulgaris, *resemble the colour of butter and egg yolks. Hence another common name for them: Butter-and-eggs.*

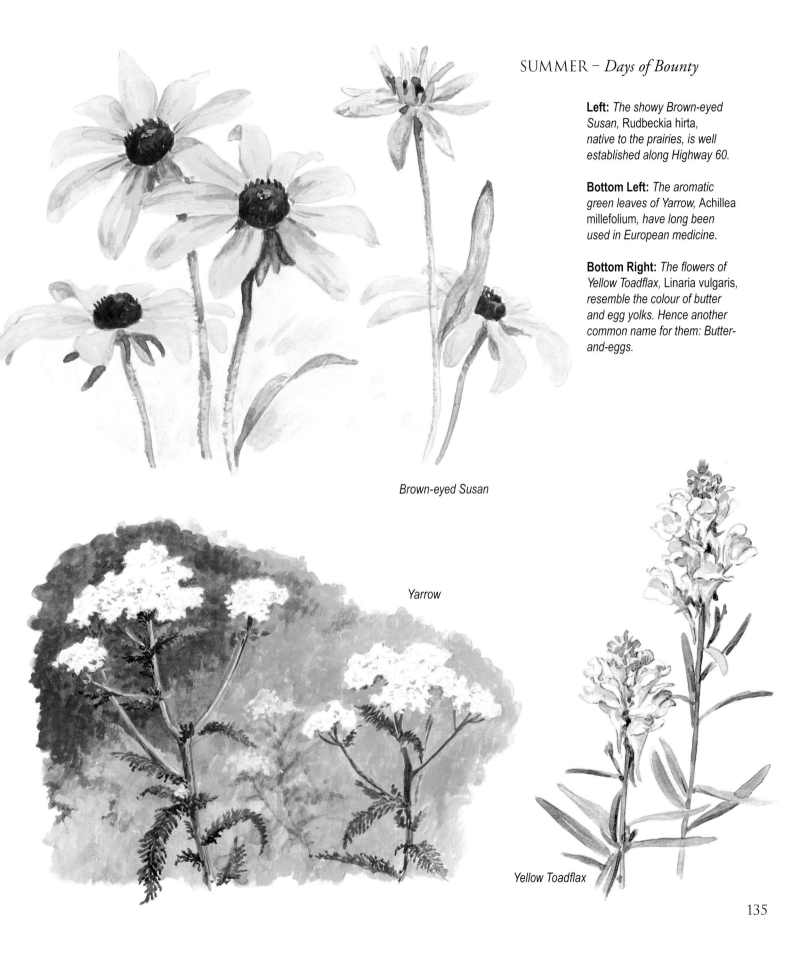

*Brown-eyed Susan*

*Yarrow*

*Yellow Toadflax*

# ALGONQUIN PARK: A PORTRAIT

## SUMMER – *Days of Bounty*

Among the many non-native flowers in Algonquin Park and in North America, is the European **Ox-eye Daisy**, *Chrysanthemum leucanthemum*. This is an old immigrant, one of the first in the New World, and was first noted some four hundred years ago.

Ox-eye Daisies, *painting by Jan Rinik*

# ALGONQUIN PARK: A PORTRAIT

The elegant and graceful **White-tailed Deer**, *Odocoileus virginianus*, is a well-known and natural part of the scenery in the eastern woods of the US and southern Canada. In Algonquin, however it is actually a newcomer that, for a time, greatly perturbed the natural ecosystem.

Originally occurring no further north than extreme southern Ontario, deer populations started to expand north in the 1800s, probably reaching what is now Algonquin in the 1870s, and continuing on well into northern Ontario. For much of the last century, deer in Algonquin maintained extraordinarily high numbers, thought by some to be in the range of thiry to one-hundred thousand. Signs of their presence were everywhere. Most lakes had distinct 'browse-lines', indicating the height to which deer had removed all shoreline cedar foliage within their reach. Regeneration of formerly abundant plant and tree species was essentially wiped out — from Ontario's provincial flower, the White Trillium (see page 39), to Canada Yew,

*Twig of a young Sugar Maple — an excellent food source for the deer*

*White-tailed Deer at Amikeus Lake*

White Cedar, Yellow Birch and Eastern Hemlock. Even the mighty moose dwindled into extreme rarity because it was being sickened and killed by a parasitic brainworm brought in by the deer.

*For most of the day White-tailed Deer remain hidden from human eyes. The best chances of spotting them are at the forest edges and near water. If you stay quiet and motionless you may be of interest to the deer as they have a strong sense of curiosity. But make the slightest movement and the deer may suddenly make a loud, short and penetrating exhalation and bound away with its snowy-white tail held upright.*

Today, of course, White-tailed Deer are relatively rare in Algonquin and the signs of their former impact have mostly disappeared. The once-threatened tree species are now regenerating satisfactorily and the moose has returned to its former status as the large animal most often seen by park visitors. But that raises important questions — four of them, in fact: Why have deer declined so much? What allowed them to reach such high numbers in Algonquin Park in the 1900s? Why did they expand northwards in the 1800s? and, most important of all, what prevented them from expanding northwards in the first place (back in the pre-settlement era)? The traditional answer to the last question is that pre-settlement deer simply could not survive in what later became their northern expansion area; the original, old-growth forests were so shady and so little sunlight reached the forest floor that there supposedly was not enough shrubby growth on the forest floor for deer to live on. We now know, however, that this explanation cannot be true; the northern area from which deer were originally excluded had been perfectly capable all along of supporting deer. We know that because other places (particularly Quebec's Anticosti Island) with even poorer food resources and worse winters than the original northern deer exclusion area have supported high numbers of deer for over a century. After all, if deer are not excluded from Anticosti Island with its bad food and climate conditions, there is no way that they were excluded in the Algonquin area, which had better conditions.

Instead, there is now good reason to believe that wolves (absent from Anticosti Island) were what kept mainland deer from expanding north into areas including what later became Algonquin Park. Ordinarily, wolves would not have been able to do this. They could have lowered the numbers of deer, but would not normally have lowered them so far as to start running out of food themselves. What was different in the future northern expansion area was that moose were also present. When deer started to become scarce, wolves could turn to moose and sustain themselves that way — even as they finished off the last of the deer. In other words, deer were

# SUMMER – *Days of Bounty*

*Portrait of the White-tailed Deer buck (male), his footprints, and tourists feeding deer in early Algonquin Park days.*

excluded from their future expansion area, not by hostile food and climatic conditions, but by the combination of a predator (the wolf) and a suitable alternative supporting prey, namely moose. With this view in mind, the answers to our other three questions fall into place. Deer expanded north because humans started to exterminate or suppress the wolves which had been preventing deer from moving north earlier. Deer reached such high numbers in Algonquin Park in the 1900s because the suppression of wolves allowed the deer to arrive in the first place, because there was no hunting and because logging and forest fires improved the food supply for the newly arrived deer. And finally, deer have drastically declined since the 1960s in Algonquin Park because that was when protection for wolves was restored and, since moose were still present, the wolves were once again able to drive deer down towards complete local extinction. They have not completely succeeded — but only because many deer leave the park in winter and, in the modern era, there are many wolfless areas close to Algonquin that support high numbers of deer. These areas act to resupply the park with new deer as the wolves eliminate most of the ones already there.

Some people miss the days when dozens of roadside deer were a common sight in Algonquin but there can not be any doubt that that was an unnatural and ecologically damaging situation. Today's Algonquin Park, with its protected wolves, good numbers of moose and quite-rare White-tailed Deer is much closer to the original, completely natural Algonquin that prevailed before the arrival of European man.

# SUMMER – *Days of Bounty*

The deer found in Algonquin belong to the largest race of White-tailed Deer, with smaller races being found in the southern portion of this species' range. The buck may reach a shoulder height of over one meter and weigh over one hundred kilograms.

Late Summer Forest
— White-tailed Deer,
*painting by Jan Rinik*

# AUTUMN *Nature's Palette*

Autumn Gold, *painting by Martin Rinik*

# ALGONQUIN PARK: A PORTRAIT

All four distinct seasons in Algonquin have their charm and fascination, but autumn is magical. The transformation of the deciduous forest from summer green to a variety of hues, ranging from light yellow to saturated yellows, from gold to orange, and from vibrant red, scarlet and ferruginous, to brown must be one of nature's grandest miracles. Lit by the sun, the forest flames vibrant colours in front of the impressed visitor. It is no wonder that the annual display of colours in Algonquin Provincial Park attracts so many people each fall.

At the end of summer, yellow begins to appear in some treetops indicating that the glucose-producing role of leaves is approaching an end. With the first leaves shed, autumn begins. Of the park's many tree species, maples are the most flamboyant. Red Maple turns first, either to a brilliant scarlet in the case of trees with all male flowers or to a bright yellow in the case of trees with all female flowers. Two weeks later, in late September, Sugar Maple puts on the biggest and most spectacular show of all — entire hillsides clothed in gold, orange and crimson. Many other trees are beautifully coloured as well: the birch, the aspen, the beech and oaks, and even the tamarack, a conifer whose soft needles turn yellow and are shed just like the leaves of deciduous trees. The colour season in Algonquin is relatively short, reaching its peak at the end of September. By the end of October the last remaining leaves are dropped and the forest is prepared for winter.

In autumn the days are shorter, nights cooler and the north winds stronger. Wildlife will forage for food, including nuts, which form an important component of their diet, before the arrival of winter. Black Bears feed voraciously on acorns, beechnuts, berries and mushrooms. Chipmunks, squirrels and Grey Jays are busy working too as they store food to use during the winter months. The warblers have already left for their warmer southern wintering grounds, as has the Broad-winged Hawk, the park's most numerous raptor. Many other birds travel south in flocks. Loons move out in October but some individuals linger long into November or even December and may sometimes get 'frozen in' and die.

Late summer and fall is also the time of wolves. Although

AUTUMN – *Nature's Palette*

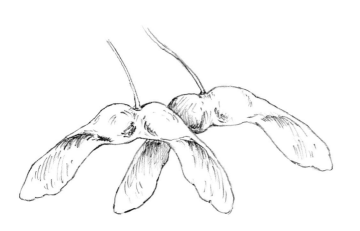

rarely seen, these iconic animals become more vocal as the pups mature. Algonquin evenings resonate with the sound of deep howls before the pack members head out from their rendezvous sites. Listening to wolves howl is very popular with the public and Algonquin Park is one of the best places on earth to hear their remarkable concerts under star-studded skies.

# ALGONQUIN PARK: A PORTRAIT

The fruit of the oak, the acorn, is widely known by its characteristic shape. Acorns are an excellent source of stored energy, which is appreciated not only by woodpeckers and jays, but also by much larger animals like the White-tailed Deer and the Black Bear.

Over sixty different oak species occur in North America but **Red Oak**, *Quercus rubra*, is the only one normally found in Algonquin. It is quite common on the park's east side, but rare on the west side, including the Highway 60 corridor. Here the species has much more localized distribution and it is mainly restricted to dry, south-facing hillsides and ridges. Yet, it can be abundant in those areas. Rusty patches on otherwise leafless hillsides betray its presence after the other trees have lost their leaves.

*Red Oak leaves and acorns*

Raccoon on a Red Oak, *painting by Jan Rinik*

*Red Oak leaves turn a beautiful gold and red in the autumn.*

149

Red-breasted Nuthatch

White-breasted Nuthatch

Brown Creeper

Many insect-eating birds live in Algonquin, but with winter coming most of them will migrate south. Only those insect-eating birds that specialize in tree-bark insects stay year round. The two species of nuthatches are small, chunky, short-tailed tree acrobats, often foraging in a head-down position on the trunk. The **Red-breasted Nuthatch**, *Sitta canadiensis*, lives in the coniferous forest. Apart from insects it favours conifer seeds, especially during the winter. It will store them for later consumption in bark crevices or in the feeding holes made by Yellow-bellied Sapsuckers. To open a spruce seed, the nuthatch will hammer it with its bill, while the seed is wedged into a crevice in tree bark.

The **White-breasted Nuthatch**, *Sitta carolinensis*, is closely associated with the hardwood forest of Algonquin. Its white face and breast, in contrast with its black crown, is distinctive and different from the broad black eye-line seen in the Red-breasted Nuthatch. Both nuthatches often join flocks of chickadees in winter.

The nuthatches are very vocal and announce their presence by loud, nasal calls. The **Brown Creeper**, *Certhia americana*, on the other hand, is an inconspicuous, well-camouflaged, tiny bird. Foraging for insects, it usually spirals upwards around tree trunks. Creepers are solitary birds most of the time, but by night family members find an appropriate hole in a tree and they will all put their heads and bodies inside, often leaving only their tails outside. By piling up on each other they retain more body heat, a necessary survival strategy on cold nights in Algonquin Park.

AUTUMN – *Nature's Palette*

*Whitefish Lake surrounded by forest as viewed from Centennial Ridges*

AUTUMN – *Nature's Palette*

Grey and Blue Jays are familiar bird species in Algonquin Park. The **Grey Jay,** *Perisoreus canadensis,* is a characteristic bird of the north, associated with spruce forests. Autumn and winter park visitors have good opportunity to meet this bird; the Grey Jay is very curious and tame and will often visit picnickers or campers and patiently wait for scraps of food. Since there is little other wildlife to see at that time of year, the experience with jays can be a trip highlight. The Grey Jay starts breeding early in March when the forests are still covered by deep snow. This species has been studied for many years in Algonquin Park; the coloured bands on the legs of this bird allow it to be individually identified by researchers.

The **Blue Jay***, Cyanocitta cristata,* is a well-known backyard bird that is also found in Algonquin. Most of the year it is noisy, with extremely loud and varied calls. The Blue Jay can imitate calls of other birds, including raptors. The beautiful cobalt-blue colour brings excitement to the autumn and winter scenery. At the end of summer and into the autumn Blue Jays collect thousands of acorns and beechnuts, transporting them in their bills to many cache sites and burying them in the ground. Grey Jays store food, berries, mushrooms, pieces of meat and some insects in spruce trees under the bark. Both species must have very good memories, since they use many cache sites to survive Algonquin's harsh winters.

**Left:** Sunset over Peck Lake, *painting by Jan Rinik*

Grey Jay

Blue Jay

*Black bear close to Mizzy Lake Trail*

There are hundreds of Black Bears in Algonquin but the chances of seeing them are usually very low. Why might this be? It is because these bears are extremely shy and avoid contact with people. Visitors are more likely to see some signs of bear presence, like piled branches resembling large birds' nests in trees, bear droppings or claw marks on beech trees.

The **American Black Bear**, *Ursus americanus*, is the second largest mammal in the park, after the moose. Large males weigh up to 150 kilograms, but there are some records of even heavier animals. Adult males tend to be about one and a half times heavier than adult females. All bears are heaviest in the autumn, before hibernation, and they weigh much less after winter when they emerge from their dens. The differences in size and weight between sexes are noticeable in young bears, usually by the time they are one year old.

# AUTUMN – *Nature's Palette*

*Portraits and foot-print of a black bear*

Bears spend most of their time foraging. Their acute sense of smell helps them locate their food sources, even from great distances. The sense of smell is actually several thousand times better than a human's and even better than that of wolves. The American Black Bears in Algonquin are forest dwellers and, as omnivorous opportunists, they feed on grass, leaves, buds, mushrooms, ants and other insects, grubs and snails, mice, squirrels and eggs, acorns and beechnuts, cherries and all kinds of berries and other fruit. They kill deer fawns and moose calves but they also eat carcasses.

In the summer and autumn bears feed on berries and on beechnuts and acorns, when available, to accumulate fat before the long winter arrives. They readily climb up trees to get acorns and beechnuts. They can sometimes be spotted while busy with this activity, offering a great sight to a quiet observer. Black bears will spend the winter in rock caves or crevices, cavities under a fallen tree, brush piles or spaces under tree roots. Some of these winter dens are surprisingly open, exposing their bodies to chilly winds and frost. With no food intake, black bears survive the Algonquin winter solely on their fat reserves, aided somewhat by their reduced body temperature.

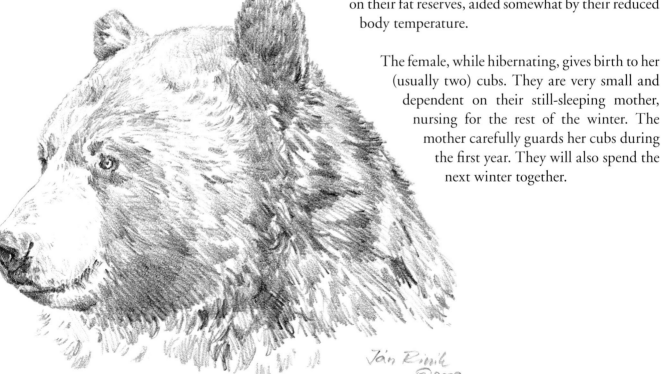

The female, while hibernating, gives birth to her (usually two) cubs. They are very small and dependent on their still-sleeping mother, nursing for the rest of the winter. The mother carefully guards her cubs during the first year. They will also spend the next winter together.

## ALGONQUIN PARK: A PORTRAIT

Blackberries and raspberries thrive in places exposed to sunshine. This can be at forest edges, clearings, along roadsides or in areas devastated by fire, windstorm or by human activities. These plants have compound and toothed leaves, and branches with thorns, often arching towards the ground. The spiny branches are called canes. Loaded with ripe berries, the canes of the **Common Blackberry**, *Rubus allegheniensis*, and of the Red Raspberry are especially attractive to Black Bears.

Common Blackberry

Peck Lake, *painting by Jan Rinik*

AUTUMN – *Nature's Palette*

*Reindeeer lichens and Largetooth Aspen leaves.*

*Hair-cap moss.*

*True Reindeer lichen (Cladonia rangiferina).*

The world of plants is extremely diverse and the Algonquin environment supports many forms, ranging from giant pines to small, inconspicuous mosses and lichens. There are 165 species of lichens and 181 species of mosses recorded in the park. Many species are hard to identify and require microscopic attention by a trained specialist.

Lichens are remarkable, often surviving in the most extreme environments. They are unique life forms, built of two separate organisms living in symbiosis: an alga, which has the ability to produce sugar from air and sunlight, and a fungus, which provides structure and extracts minerals from whatever they are growing on. In Algonquin Park, at the Lookout cliff, they cover bare rocks with their colourful crusts and diverse forms. Lichens may form large colonies and can live many hundreds of years; some can be considered the oldest living things on our planet. Most are exposed to conditions like direct sun and rain, severe winds, frost, ice and snow in winter, under which their growth is very slow. Lichens not only survive, but slowly erode hard rock surface, preparing little pockets of soil for other living organisms including a succession of plants, eventually leading to trees.

One beneficiary of lichens' activities is moss, the small green plant that plays a role in soil formation. Mosses prefer places with high humidity. Unlike other plants, they have no water-transporting system. They also lack true leaves, flowers and roots. Many species are almost identical and can only be distinguished by microscopic examination.

AUTUMN – *Nature's Palette*

Berm Lake Trail

*Growing on old wood, sandy soils, bare rocks or hanging in tufts on trunks and branches of conifers, lichens show their wide diversity of forms. Each tree species and each type of rock supports different kinds of lichen. Greyish, greenish and brownish hues are their most usual colours. Humble lichens need no soil. They thrive in places where no other plants can grow.*

# ALGONQUIN PARK: A PORTRAIT

Poplars, including the park's two species of aspens, are quick-growing hardwoods, but as they age many hollows will form in trunks and branches. Thus, they are important to animals, birds and insects as breeding and foraging sites. These species, along with White Birch, like sunshine and thrive as pioneer trees, growing on disturbed and burned-over areas. Poplars are characteristic sights along stream banks, but are also found in quite dry areas. Three species are found in Algonquin, of which the most common are the Trembling Aspen and Largetooth Aspen. They are similar in appearance with light, almost white bark. Sometimes, from a distance, they may be mistaken for White Birch.

The high nutritional value of the buds, leaves and twigs of the **Balsam Poplar**, *Populus balsamifera*, make it part of the preferred diets of deer, moose, beavers and grouse. The **Trembling Aspen**, *Populus tremuloides*, is famous for its leaves that flutter in the slightest breeze. In the autumn the leaves turn a bright yellow. Trembling Aspen is one of the beavers' most preferred trees. The **Largetooth Aspen**, *Populus grandidentata*, prefers drier, often more upland habitats.

**Above:** *The glossy, dark green leaves of a Trembling Aspen are oval in shape, with slightly serrated edges. The leaf stalks are flattened and flexible, which causes their constant movement in even a slight breeze.*

*Leaves of Largetooth Aspen are sharply cut, especially those found on older twigs. Younger branches have more oval shaped leaves.*

# AUTUMN – *Nature's Palette*

*The word tamarack has its origin in the Algonquian group of languages, along with other words like moose, skunk, moccasin and tomahawk (illustrated below), which have been adopted into English.*

The spectacular autumn palette of colours is usually associated with the broad-leaved deciduous forest. As the trees stop the process of photosynthesis, their leaves turn yellow and finally fall before winter. The **Tamarack**, *Larix laricina*, is the only conifer that does the same. Its clusters of soft needles, vivid green in summer, turn to bright yellow and gold, and then they are shed. The Tamarack is also known as the Eastern or American Larch. It is found at the edges of bogs and swamps, where the moving water provides good oxygenation for its roots, an important requirement for this species.

# AUTUMN – *Nature's Palette*

**Morning at Spruce Bog,**
*painting by Jan Rinik*

Great Blue Heron

163

# AUTUMN – *Nature's Palette*

**Below:** *The bulk of the diet of the Deer Mouse,* Peromyscus maniculatus, *consists of Sugar Maple seeds, especially in the autumn. They store masses of them and will survive the long winter eating them. The numbers of Deer Mice in Algonquin fluctuate with good and weak crops of Sugar Maple seeds.*

Brilliant colours of Algonquin hardwood forest in the autumn attract thousands of visitors. The highlight of the Algonquin year is the end of September and the beginning of October, as the green leaves turn to colours ranging from yellow and gold to deep red and scarlet. Without a doubt it is one of our planet's most marvellous sights and an unforgettable spectacle. The Sugar Maple the most abundant of the tree species that undergo such seasonal changes.

The Sugar Maple–dominated hardwood forests cover more than half of the park's area. They are one of the most characteristic natural history features in Algonquin. They can be seen in the Hardwood Lookout Trail, as well as in many other areas along the Highway 60 corridor on the west side of the park. Sugar Maple stands are supported by soil left by the glaciers, the gravely 'glacial till' that retains moisture, providing ideal conditions for the growth of maples.

**Left and Right:** *Sugar Maple trees in their autumn beauty*

# ALGONQUIN PARK: A PORTRAIT

There are five native maple species in Algonquin of which the **Sugar Maple**, *Acer saccharum*, is by far the most dominant tree in the west-side highlands. In many west-side Algonquin forests Sugar Maples account for over 95 per cent of all the trees present. Other trees, such as Beech, Yellow Birch, Red Maple, Hemlock and White Pine, or rarely Basswood and Black Cherry, account for the remainder. Sugar Maple trunks are straight, sometimes up to thirty metres tall, and their canopy of green leaves keeps the forest floor in deep shade all summer. Sugar Maple's five-lobed leaves turn vivid yellow, orange or bright red in the autumn. The seeds, being equipped with wings and joined in pairs, will slowly descend after they mature in autumn, spinning in the wind and often reaching areas well away from their parent trees.

*Sugar Maple*

AUTUMN – *Nature's Palette*

**Right:** *Twig of a Silver Maple*
**Below:** *Brightly coloured leaves of Red Maple, which are yellow in female trees and scarlet red in male trees*

Silver Maple

Red Maple leaves

The **Red Maples**, *Acer rubrum*, are also conspicuous in their autumn colours. Trees that have predominantly male flowers have leaves that turn bright red and scarlet, while predominantly female tree leaves turn into beautiful yellow. Flowers of both sexes on Red Maples are red as well. The leaves of the majestic **Silver Maple**, *Acer saccharinum*, are deeply lobed with silver coating on their underside. Though well known from city parks and streets, it is quite rare in Algonquin. It can be seen here in its natural habitat, inundated floodplains along the Oxtongue River, Petawawa River, Barron River and a few other places. Hybrids between Silver and Red Maples that share characteristics of both parent species are even more common than the Silver Maple.

# ALGONQUIN PARK: A PORTRAIT

AUTUMN – *Nature's Palette*

Red Maples and White Birches, *painting by Jan Rinik*

# ALGONQUIN PARK: A PORTRAIT

Park visitors are keen on seeing wildlife and the **Red Fox**, *Vulpes vulpes*, is one animal many people get to see as it is often active in daylight hours. In addition, some lose their fear of humans and will approach campers quite closely in the expectation of being fed. The Red Fox is a handsome animal with nice rusty-coloured fur, black legs and a long, white-tipped tail. Rusty-coated Red Foxes are unmistakable and by far the most numerous in Algonquin, compared to quite-scarce greyish-brown and black varieties of the same species.

# AUTUMN – *Nature's Palette*

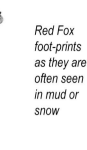

*Red Fox foot-prints as they are often seen in mud or snow*

The Red Fox in Algonquin is an opportunistic and adaptable forager, eating any food that is locally available. It hunts solitarily, unlike wolves, where the whole family participates. Red foxes feed on small mammals including rodents, chipmunks and hares, small birds and their eggs, and on larger insects and some berries. Their senses are excellent, especially hearing and smell, which helps them locate prey. Mice make up the main component of their diet.

Foxes have a wonderful hunting technique, catching mice with a characteristic high jump. They wait patiently and then leap up in the air, landing forepaws first on the prey before delivering a killing bite. In the winter, when the opportunity arises, Red Foxes readily feed on remains of wolf kills. While scavenging, they are occasionally killed by wolves. Foxes are particularly vulnerable to rabies, but in Algonquin this disease is very rare.

# ALGONQUIN PARK: A PORTRAIT

Lynx

The Algonquin mammal list includes three cat species, although it is unlikely that any is a permanent resident in the park. The **Canada Lynx**, *Lynx canadensis*, is a boreal forest dweller. With long legs and wide paws, it is well adapted for travelling in deep snow. Its main diet is the Snowshoe Hare. The lynx's natural range stretches further north, with only a few sightings reported in Algonquin and surrounding area.

AUTUMN – *Nature's Palette*

*Eastern Cougar*

The **Bobcat**, *Lynx rufus*, is similar in appearance to the lynx, but more reddish, spotted and smaller. It is typically an inhabitant of southern forests — being less adapted to deep-snow conditions — where it preys on deer fawns, rabbits and birds. The bobcat is an excellent tree climber. Both the lynx and the bobcat have very short tails.

The **Eastern Cougar**, *Felis concolor*, is a tawny brown cat with an extremely long tail. It was present in the eastern forests and may have lived in the Algonquin area when White-tailed Deer were abundant. However, today the cougar is not known to inhabit the park. There are many alleged sightings from all over Ontario and southern Quebec, including close to Algonquin, but since cougars are often kept in captivity the animals reported may be escaped or released captives. That being said, any and all sightings of these three cat species should be reported to park staff.

**Below and Opposite:** *Both Bobcat and the Canada Lynx are short-tailed cats with body length of seventy to one hundred centimetres plus ten centimetres for a tail, weighing thirteen to fifteen kilograms.*
**Above:** *The Eastern Cougar has up to a metre long tail and its general body weight ranges from forty to seventy kilograms.*

*Bobcat*

The long wolf howl, piercing the night silence, terrified people for hundreds of years. As elusive and mysterious threats to our livestock, wolves have long had a very bad reputation. Even in Algonquin Park, the howls originally brought a clear message: wolves were here, and this presence was not to be tolerated. Park legislation expressly called for their extermination using guns, traps, poison, snaring and every other possible method. Even as late as 1959 park rangers were paid a bounty for every wolf they killed. But finally the destruction stopped to make way for one of the earliest scientific investigations of wolf behaviour and ecology.

Since the 1960s, Algonquin Park has been a centre of wolf research and thanks to that it has helped to establish a better understanding of this enigmatic predator. Although the knowledge is still far from being complete, it is already good enough to shed a new light on wolves and their role as top

*Algonquin Provincial Park is world famous as a place to hear the wolf howl.*

# AUTUMN – *Nature's Palette*

*It seems that the northern limit of Eastern Wolves coincides with that of White-tailed Deer. Certainly, wolves in Algonquin prefer White-tailed Deer as a food source, with beaver and the much more dangerous and hard-to-kill moose available as secondary choices. Wolves live and hunt in families, usually up to five or six members. Their territories are about 150 to 200 square kilometres, and all together the number of wolves in Algonquin is estimated at a maximum of 250.*

predators, not only in the Algonquin area but also in their large natural range all over the northern hemisphere.

**Eastern Wolves**, *Canis lycaon*, of central Ontario, especially Algonquin Park, were until recently considered a smaller race of the Grey Wolf, the widespread canid that shows many geographical variations. But, in the 1990s, DNA studies revealed that Algonquin wolves are genetically more related to the rare Red Wolf of the southern US than to the Grey Wolf of the north. This makes the wolves of Algonquin even more unique and noteworthy.

Today, Algonquin's starry nights echo with haunting wolf howls once again. Their message is loud and clear and much different from that of past centuries: the presence of a most admired, fascinating and appreciated animal. Without a doubt, the wolf is on the most wanted list for those thousands of visitors who take part each year in public wolf-howl events conducted by Algonquin naturalists. For many of these visitors, this is the most important reason to visit the park. It seems that the change of reputation of wolves has contributed, at least locally, to the preservation of one of the world's most intelligent and charismatic animals.

# AUTUMN – *Nature's Palette*

Birches in Gold,
*painting by Jan Rinik*

Thanks to its unmistakable white bark streaked with black, the White Birch is an easy identified and well-known tree. In Algonquin Park two birch species are common. The creamy white bark is characteristic of **White Birch**, *Betula papyrifera*, which is also known as Paper Birch or Canoe Birch. The White Birch quickly colonizes disturbed land and begins a new forest at locations devastated by logging or forest fires.

The **Yellow Birch**, *Betula alleghaniensis*, is taller and larger, but less conspicuous due to its muted bark colour. It grows in damp areas of mature hardwood forests, which is very different from White Birch. Its seedlings are very favoured by White-tailed Deer.

# ALGONQUIN PARK: A PORTRAIT

The **American Woodcock**, *Scolopax minor*, is relatively unknown to park visitors, perhaps because it is active at dusk and night, and its cinnamon-coloured plumage blends with its surroundings. Woodcocks are quite common in the park from snowmelt to the first days of winter. They are chunky birds with long, straight bills used to probe for earthworms and other soil invertebrates well below the forest floor's surface. Moist woodlands and alder thickets are their preferred habitat for breeding.

American Woodcock

*The first winter frost at Wolf Howl Pond, Mizzy Lake Trail*

AUTUMN – *Nature's Palette*

# ALGONQUIN PARK: A PORTRAIT

*Morning mist at Tea Lake*

When Algonquin lakes freeze over in December, life is no longer possible for the park's fish-eating birds. Indeed, well before freeze-up, herons, ospreys, loons, kingfishers and mergansers have all left for the south.

Mergansers are well adapted to their way of life. They have a slightly hooked bill, serrated along its edges, and perfectly suited to grip small fish such as minnows and others. They locate their prey by having their heads submerged in water and scanning. After spotting a fish they immediately dive and pursue it underwater by strong strokes of their large webbed feet, holding their wings folded close to the body.

Two kinds of mergansers are common in Algonquin. The one seen most often is the **Common Merganser**, *Mergus merganser*. Males are beautiful white birds with dark green heads, while females have rusty brown heads. They prefer deep lakes

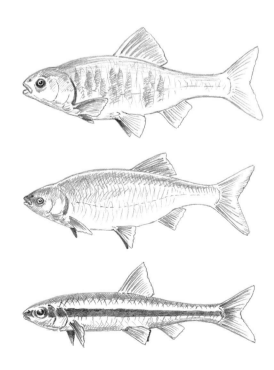

# AUTUMN – *Nature's Palette*

with cold water and rocky shores. Common Mergansers tend to fish in rivers and smaller lakes — they may be attacked and killed if they attempt to fish in areas used by loons. After mating in June, males depart to the north, leaving females incubating eggs and rearing chicks. During the summer all Common Mergansers seen in the park are females or young birds. Mergansers usually nest in tree cavities, and when the young hatch they leave the nest — jumping one by one down from the hole, sometimes from a sizeable height.

The male **Hooded Merganser**, *Lophodytes cucullatus*, is an elegant bird with a large crest and with white, black and brown plumage. Females are a dull, smoky greyish-brown. In any case, Hooded Mergansers live in small out-of-the way ponds and tend not to be noticed by summer visitors. They are most visible right after their arrival to the park, in early April, and then again after breeding in autumn and winter. Only the first ice cover on the lakes forces them to leave the park.

**Opposite page top:** *The Common Shiner,* Luxilus cornutus
**Opposite page middle:** *The Golden Shiner,* Notemigonus crysoleucas
**Opposite page bottom:** *The Blacknose Shiner,* Notropis heterolepis. *All are widely distributed in Algonquin beaver ponds, lakes and rivers with abundant aquatic vegetation. They are small fish, 7 to 10 cm long, with the Common Shiner rarely reaching 20 cm. These three species belong to the minnow family (of which Algonquin Park has 18 species).*

Common Merganser male

Common Merganser female

Hooded Merganser

# ALGONQUIN PARK: A PORTRAIT

Many fish species in Algonquin lakes and rivers have their spawning season in late spring and in summer, and their fry (the newly hatched young) develop a few days later. Before the arrival of winter they are already large enough to survive it. The spawning of the **Brook Trout**, *Salvelinus fontinalis*, follows a different schedule. With the last leaves of maple trees fluttering in the air, the annual trout spawning season is coming. At the beginning of November male Brook Trout are at their most brilliant. They have olive green backs, red spots with bluish halos around them and red sides with black edges underneath, in contrast to their white belly. The deep orange eggs are laid in spawning beds in shallow waters. Buried in gravel by the female, they are hidden from predators over the winter. If all the requirements are fulfilled, the young trout will emerge and start their own new lives the following spring.

Spawning Season — Brook Trout, *painting by Martin Rinik*

AUTUMN – *Nature's Palette*

183

# WINTER *Time to Persist*

Mew Lake in Winter, *painting by Jan Rinik*

# ALGONQUIN PARK: A PORTRAIT

Since many visitors think of Algonquin as being 'up north', it is appropriate that the season that brings true northern character to the park is winter. Most years, from December to March, snow in the park offers good cross-country skiing and snowshoeing conditions and many visitors will time their trips for that period of the year.

Winters in Algonquin are long and hostile. By the end of autumn the last leaves have fallen and the forest scenery is very different. An overnight snowfall can transform murky grey forest into a brilliant white, snow-coated realm. The protective blanket of snow shields the life beneath as lashing winds and bitter frosts descend on Algonquin. Persistence without adaptation is out the question. Trees enter a state of dormancy where most of their activities are shut down. Some animals hibernate deep in their dens, others like beavers and Grey Jays subsist on food stored the previous autumn. If the snow is particularly deep, deer and moose can have hard times obtaining food and may fall prey to wolves. Their carcasses attract ravens, eagles, foxes, fishers and others. Wolves themselves may have a difficult time and the lack of food may lead to a high death rate among pack members.

Ice cover on lakes and rivers isolates life underneath from the outside world. Water that is rich in oxygen and nutrients supports fish and countless small creatures who survive in the relatively warm environment of 4ºC, while the outside temperature may drop down to minus 40ºC.

# WINTER – *Time to Persist*

Many bird species have migrated south before the arrival of winter. Those species that stay in Algonquin can suffer from food shortages and cold temperatures. Some winter birds, like crossbills, are nomadic, and their flocks search large areas of forest for pine or spruce cones to survive. Winter puts every creature on trial, small or large. In their struggles to persist they either survive winter or die. Again and again, from year to year, nature repeats its miraculous seasonal cycle of life and death in Algonquin Park.

# ALGONQUIN PARK: A PORTRAIT

*Black-capped Chickadee*

*Dark-eyed Junco*

*Downy Woodpecker*

Several small birds remain in the park when winter snows cover the land. They are usually not evenly spread, as they might be in summer, but rather they tend to congregate in particular spots where food is available. Such gatherings can consist of many birds, and one of the most numerous is the **Black-capped Chickadee**, *Poecile atricapillus*. Chickadees are small and familiar to those who feed birds in backyards and parks. As they get used to people, chickadees will often come to pick up seed right from the palm of your hand. Less familiar to visitors are the very similar Boreal Chickadees. This species has a brown cap instead of black and typically is not found south of Algonquin.

Regularly seen in summer and only rarely present in winter is the **Dark-eyed Junco**, *Junco hyemalis*. These simply coloured birds have white outer tail feathers.

**Downy Woodpeckers**, *Picoides pubescens*, are often seen browsing small tree branches and weed stems for insects. Sharing the same black and white colouration, Downy Woodpeckers are similar to the larger Hairy Woodpecker.

WINTER – *Time to Persist*

During the long nights of winter small birds find shelter in dense conifers where they are safer from predators and better protected from the cold. During the day, owls (their potential predators) are largely harmless, and if one is spotted small birds may actually mob it in an effort to make it move out of their territories. However, owls and small hawks are scarce during harsh winters in Algonquin, and the biggest problem remains very low temperature. Snow is an effective insulator and by roosting under conifer branches heavily laden with snow, these small birds lower their radiant heat loss. Their bodies also have the special adaptation of saving metabolic energy by dropping their body temperature while inactive.

**Left:** *The Barred Owl,* Strix varia, *is a winter resident in Algonquin Park.*

# ALGONQUIN PARK: A PORTRAIT

Two species of crossbills are irregular winter visitors to Algonquin Park. They are food specialists, feeding on seeds of conifers such as spruce, tamarack and pine. As the cone crops fluctuate from year to year in the park, as well as in other regions of boreal forest, these highly nomadic birds may appear in thousands in one year and be completely absent in another, all in response to their food availability.

While feeding on hard cones, both the **Red Crossbill**, *Loxia curvirostra*, and the **White-winged Crossbill**, *Loxia leucoptera*, use a special technique. They insert their bill between the scales of the cone, sideways lever apart the bracts with their crossed mandibles and remove the seed with their specialized scoop-like tongue. This extraordinary adaptation is very efficient. A crossbill can consume up to three thousand seeds per day.

If their food supply is especially good, crossbills may build their nests and raise young even in the middle of winter. With coming spring, most cones open and shed their seed, and food availability is reduced. Nomadic crossbills then usually leave the area in search of other regions.

**Above:** *Red Crossbill exploring a White Pine cone. It is an acrobatic forager, often hanging from cones and easily fluttering from twig to twig. Its massive bill is adapted to large pine cones and the species can be seen more often in the east side of the park.*

**Opposite:** *Foraging flock of White-winged Crossbills. With their slender bills, these birds exploit smaller cones of conifers, such as Tamarack and spruce. White Spruce and Black Spruce are the most important sources of food for the species in the park, and many thousands of birds may gather in spruce stands in a good crop year.*

**Right:** *Among other finches, often seen in winter, are Evening Grosbeaks,* Hesperiphona vespertina. *Small flocks can be seen along the highway picking grit to help digest seeds.*

# ALGONQUIN PARK: A PORTRAIT

# WINTER – *Time to Persist*

Animal tracks can reveal animal activities and distribution. Many mammal species are nocturnal and rarely sighted; the variety of tracks recorded in snow can be surprising.

The park is a refuge for animals, like those prized for their fur. Beaver, marten, fisher and mink were heavily trapped in Ontario, even in Algonquin before it became a park. Once their numbers recovered, beavers were live-trapped and shipped to the US, Europe and Asia to restock where they had been extirpated.

*Trapper and a Red Fox*

Red Foxes in winter, *painting by Jan Rinik*

# ALGONQUIN PARK: A PORTRAIT

Winters in Algonquin are harsh and food is scarce. Two species of large eagles can be seen at that long period of hardship and they are both attracted to carcasses of wolf kills. In Algonquin Park, the large, hoofed mammals — the White-tailed Deer and the Moose — are the most common winter diet of Eastern Wolves and their scraps are often left on frozen lakes.

The **Bald Eagle**, *Haliaeetus leucocephalus*, is a majestic bird with a wingspan of over two metres. Adults, as seen in the picture, have a pure white head and tail, with a dark brown, almost black, body. These birds have a massive hooked beak. From November until March, when food is available in the form of leftovers, the Bald Eagle can be seen scavenging.

From time to time the **Golden Eagle**, *Aquila chrysaetos*, appears in the park. Contrary to the Bald Eagle, that now occasionally breeds in the park, the Golden Eagle never appears here except in winter. The Golden Eagles are spectacular raptors with powerful talons. The juvenile birds have a noticeable white base to the tail and a dark brown terminal band. The nearest breeding grounds of the Golden Eagle, and the probable origin of birds that appear in Algonquin, are in northern Quebec.

**Above:** *Golden Eagle flying*

**Right:** *Bald Eagle on a pine tree*

*Bald Eagle and Ravens on a deer carcass*

# ALGONQUIN PARK: A PORTRAIT

The **Northern Goshawk,** *Accipiter gentilis,* is present year round in Algonquin, but escapes attention due to its secretive habits. It preys on mammals and birds the size of the Snowshoe Hare and Ruffed Grouse.

*Goshawk with a Grey Jay*

Eastern Wolves,
a family pack
searching for prey,
*painting by*
*Jan Rinik*

# ALGONQUIN PARK: A PORTRAIT

Our journey through the four seasons of Algonquin Park is now over. On the previous pages we could only take a quick look at a select few of the park's diverse plants and animals. It is our hope that this account will encourage every reader and nature lover to set out on his or her own fascinating journeys of discovery through the beautiful Algonquin environment. If this occurs, then our goal as authors will be fulfilled.

The cycle of life in the park has always been full of changes. There are changes in the seasons as they repeat themselves year after year, being similar but not identical. There are also more dramatic changes in the relatively short history of the park. From well before its establishment in 1893 and up to the present, Algonquin Park has been the scene of important logging activity. Wolves, for which today's Algonquin is so famous, were heavily persecuted. White-tailed Deer reached enormous numbers, followed by a drop in moose. Later, in the late 1960s and early 1970s, deer collapsed and moose rose again. The moose became the park's most visible big animal, and chances of seeing White-tailed Deer, once plentiful and tame, are low. Algonquin is the same as, yet subtly different than, it was in the past. But thanks to better understanding of ecology, the park nowadays is probably more natural than it was at the time of its establishment.

People also change a lot from generation to generation. Values change and different attitudes are adopted. Therefore we should keep one thing at the forefront of our minds: Algonquin is a special and unique place, offering all of us much more than that which is found in the pages of this book. Let us do everything within our abilities to ensure that our children and grandchildren are able to have the same experiences while visiting Algonquin Park in the decades and centuries to come. Let Ontario's Algonquin Provincial Park be a living, breathing repository of beautiful Canadian nature, a place where everyone will always be able to enrich their lives with its beauty and harmony.

**Opposite:** *Lake of Two Rivers*

**Below:** *The River Otter,* Lontra canadensis, *is active and mobile year round. Otters hunt for fish, their main diet, but also for crayfish or amphibians. In winter they cover long distances under the ice in their search for food. Otters also travel over snow, often running and sliding, leaving characteristic trails.*

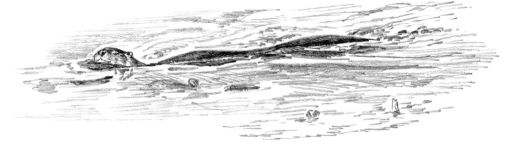

# BIBLIOGRAPHY

Alderfer, J., ed. 2006. *National Geographic Complete Birds of North America.* National Geographic, Washington, D.C. 664 pp.

American Ornithologists' Union. 1998. *Check-list of North American Birds.* 7th Edition. American Ornithologists' Union, Washington, D.C.

Backhouse, F. 2005. *Woodpeckers of North America.* Firefly Books, Richmond Hill, Ontario. 232 pp.

Beer, A.J. & Morris, P. 2004. *Encyclopedia of North American Mammals.* Thunder Bay Press, San Diego. 288 pp.

Bellrose, F. C. 1976. *Ducks, Geese and Swans of North America.* Stackpole Books, Harrisburg, Pennsylvania. 544 pp.

Benkman, C.W. 1992. White-winged Crossbill. In *The Birds of North America*, No. 27 (Poole, A., Stettenheim, P. & Gill, F., eds.). The Academy of Natural Sciences, Philadelphia, and The American Ornithologists' Union, Washington, D.C. 20 pp.

Bice, R. 1980. *Along the Trail with Ralph Bice in Algonquin Park.* Consolidated Amethyst, Scarborough. 159 pp.

Bland, J. 1971. *Forests of Lilliput: The Realm of Mosses and Lichens.* Prentice-Hall, Englewood Cliffs, New Jersey. 210 pp.

Bremness, L. 1994. *The Complete Book of Herbs.* Viking Studio, Dorling Kindersley Book, New York, London. 288 pp.

Brooks, R.J., Strickland, D. & Rutter, R.J. 2003. *Reptiles and Amphibians of Algonquin Provincial Park.* The Friends of Algonquin Park, Whitney, Ontario. 48 pp.

Brown, G.W., Harman, E. & Jeanneret, M. 1960. *Canada in North America to 1800.* Copp Clark Publishing, Toronto. 369 pp.

Cade, T.J. & Burnham, W., eds. 2003. *Return of the Peregrine: A North American Saga of Tenacity and Teamwork.* The Peregrine Fund, Boise, Idaho. 394 pp.

Cafferty, S., ed. 2005. *Firefly Encyclopedia of Trees.* Firefly Books, Buffalo, Richmind Hill, London. 288 pp.

Carpentier, G. 1987. *The Mammals of Peterborough County.* Peterborough Field Naturalists', Orchid Press, Peterborough, Ontario. 125 pp.

Chambers, B., Legasy, K. & Bentley, C.V. 1996. *Forest Plants of Central Ontario.* Lone Pine Publishing, Edmonton, Vancouver, Renton. 448 pp.

Chesser, R.T. *et al.* 2011. Fifty-second supplement to the American Ornithologists' Union Check-list of North American Birds. *Auk* 128: 600–613.

Collingwood, G.H., Brush, W.D. & Butcher, D. 1974. *Knowing Your Trees.* The American Forestry Association, Washington, D.C. 375 pp.

Cook, S.J., Norris, D.R. & Theberge, J.B. 1999. Spatial dynamics of a migratory wolf population in winter, south-central Ontario (1990–1995). *Canadian Journal of Zoology* 77: 1740–1750.

Crins, W.J., Blaney, C.S. & Brunton, D.F. 1998. *Checklist of the Vascular Plants of Algonquin Provincial Park.* Algonquin Park Technical Bulletin No.2. The Friends of Algonquin Park, Whitney, Ontario. 5 pp.

Crins, W.J. & Darbyshire, S. 1993. *Checklist of the Bryophytes of Algonquin Provincial Park.* Algonquin Park Technical Bulletin No.4. The Friends of Algonquin Park, Whitney, Ontario. 34 pp.

del Hoyo, J., Elliott, A. & Sargatal, J., eds. 1992. *Handbook of the Birds of the World.* Vol. 1. Ostrich to Ducks. Lynx Edicions, Barcelona. 696 pp.

del Hoyo, J., Elliott, A. & Sargatal, J., eds. 1994. *Handbook of the Birds of the World.* Vol. 2. New World Vultures to Guineafowl. Lynx Edicions, Barcelona. 638 pp.

del Hoyo, J., Elliott, A. & Christie, D. A., eds. 2010. *Handbook of the Birds of the World.* Vol. 15. Weavers to New World Warblers. Lynx Edicions, Barcelona. 879 pp.

Dickinson, T., Metsger, D., Bull, J. & Dickinson, R. 2004. *The ROM Field Guide to Wildflowers of Ontario.* Royal Ontario Museum, Toronto. 416 pp.

Dickson, H.L. & Crins, W.J. 1993. *Checklist of the Lichens of Algonquin Provincial Park.* Algonquin Park Technical Bulletin No.7. The Friends of Algonquin Park, Whitney, Ontario. 2 pp.

# BIBLIOGRAPHY

Dunbar, D. 1991. *The Outdoor Traveler's Guide Canada.* Stewart, Tabori & Chang, New York. 400 pp.

Farb, P. 1968. *The Land and Wildlife of North America.* Time-Life Books, New York. 200 pp.

Farb, P. 1970. *The Forest.* Time-Life Books, New York. 192 pp.

Farrar, J.L. 1995. *Trees in Canada.* Fitzhenry & Whiteside, Markham, Canadian Forest Service, Ottawa. 502 pp.

Finlay, J.C. ed. 1984. *A Bird-Finding Guide to Canada.* Hurtig Publishers, Edmonton. 387 pp.

Forbes, G.J. & Theberge, J.B. 1992. Importance of scavenging on moose by wolves in Algonquin Park, Ontario. *Alces* 28: 235–241.

Forbes, G.J. & Theberge, J.B. 1992. Response by wolves to prey variation in central Ontario. *Canadian Journal of Zoology* 74: 1511–1520.

Forsyth, A. 1985. *Mammals of the Canadian Wild.* Camden House Publishing. Camden East, Ontario. 351 pp.

Forsyth, A. 1999. *Mammals of North America.* Temperate and Arctic Regions. Firefly Books, Buffalo, New York. 350 pp.

Godfrey, W.E. 1986. *The Birds of Canada.* National Museum of Natural Sciences, National Museums of Canada, Ottawa. 595 pp.

Goller, C. 1984. *Algonkian Hunters of the Eastern Woodlands.* Grolier Limited, Canada. 48 pp.

Goodwin, C.E. 1995. *A Bird-Finding Guide to Ontario.* University of Toronto Press, Toronto, Buffalo, London. 477 pp.

Grady, W. 2007. *The Great Lakes. The Natural History of a Changing Region.* Greystone Books, Vancouver, Toronto, Berkeley. 352 pp.

Grewal, S.K., Wilson, P.J., Kung, T.K., Shami, K., Theberge, M.T., Theberge, J.B. & White, B.N. 2004. A genetic assessment of the Eastern Wolf (*Canis lycaon*) in Algonquin Provincial Park. *Journal of Mammalogy* 85: 625–632.

Hall, D., ed. 2004. *Encyclopedia of North American Birds.* Thunder Bay Press, San Diego. 288 pp.

Harrison, M. & Thompson, P. 1999. *Explore Canada. The Adventurer's Guide.* Key Porter Books, Toronto. 301 pp.

Henry, J.D. 2002. *Canada's Boreal Forest.* Smithsonian Institution Press, Washington, London. 176 pp.

Hepp, G.R. & Bellrose, F.C. 1995. Wood Duck. In *The Birds of North America*, No. 169 (Poole, A. & Gill, F.,eds.). The Academy of Natural Sciences, Philadelphia, and The American Ornithologists' Union, Washington, D.C. 24 pp.

Horsley, S.B., Stout, S.L. & deCalesta, D.S. 2003. White-tailed deer impact on the vegetation dynamics of a northern hardwood forest. *Ecological Applications* 13: 98–118.

Hunter, D. 2007. The Ghost Cat. Ontario Nature, Federation of Ontario Naturalists, Toronto. *ON Nature* Winter 2006/2007: pp 18–23, 36.

Huot, J. 1974. Winter Habitat of White-tailed Deer at Thirty-one Mile Lake, Quebec. *The Canadian Field-Naturalist* 88: 293–301.

Jones, C.D. 2003. *Checklist and Seasonal Status of the Butterflies of Algonquin Provincial Park.* Algonquin Park Technical Bulletin No.1. The Friends of Algonquin Park, Whitney, Ontario. 21 pp.

Jones, C.D., Kingsley, A., Burke, P. & Holder, M. 2008. *Field Guide to the Dragonflies and Damselflies of Algonquin Provincial Park and the Surrounding Area.* The Friends of Algonquin Park, Whitney, Ontario. 263 pp.

Jones, D. 1999. *North American Wildlife.* Prospero Books, Vancouver, Toronto, New York. 304 pp.

Karstad, A. 1979. *Aleta Karstad's Canadian Nature Notebook.* McGraw-Hill Ryerson Limited, Toronto, Montreal. 144 pp.

Kelly, R.W. 1986. *The Hurons: Corn Planters of the Eastern Woodlands.* Nelson Canada, Scarborough. 84 pp.

Kershaw, L. 2001. *Trees of Ontario. Including Tall Shrubs.* Lone Pine Publishing, Edmonton, Renton. 240 pp.

Ketchum, R.M. 1970. *The Secret Life of the Forest.* American Heritage Press, New York. 111 pp.

Little, E.L., Jr. 1979. *Forest Trees of the United States and Canada, and How to Identify Them.* Dover Publications, New York. 71 pp.

Livesey, R. & Smith, A.G. 2005. *The Fur Traders.* Fitzhenry & Whiteside, Markham, Ontario, Allston, Massachusetts. 92 pp.

Lynch, W. 2007. *Owls of the United States and Canada.* UBC Press, Vancouver. 242 pp.

MacKay, R. 2002. *A Chronology of Algonquin Park History.* Algonquin Park Technical Bulletin No.8. The Friends of Algonquin Park, Whitney, Ontario. 32 pp.

Mandrak, N.E. & Crossman, E.J. 2003. *Fishes of Algonquin Provincial Park.* The Friends of Algonquin Park, Whitney, Ontario. 40 pp.

Marshall, S. 1997. *Insects of Algonquin Provincial Park.* The Friends of Algonquin Park, Whitney, Ontario. 48 pp.

Maybank, B. & Mertz, P. 2001. *The National Parks and other Wild Places of Canada.* Barron's Educational Series, Inc. Hauppauge, New York. 176 pp.

McCormick, J. 1966. *The Life of the Forest.* McGraw-Hill, New York, Toronto, London. 232 pp.

McIntyre, J.W. & Barr, J.F. Common Loon. In *The Birds of North America*, No. 313 (Poole, A. & Gill, F., eds.). The Academy of Natural Sciences, Philadelphia, and The American Ornithologists' Union, Washington, D.C. 32 pp.

McIvor, L. 2003. *Head-of-the-Lake Pocket Nature Guide.* Hamilton Naturalists' Club, Hamilton. 106 pp.

McNicholl, M.K. & Cranmer-Byng, J.L. eds. 1994. *Ornithology in Ontario.* Ontario Field Ornithologists Special Publication No. 1, Hawk Owl Publishing, Whitby. 400 pp.

National Geographic Society. 1989. *Field Guide to the Birds of North America.* Washington, D. C. 464 pp.

National Geographic Society. 1998. *Exploring Canada's Spectacular National Parks.* Washington, D.C. 200 pp.

Newmaster, S.G., Harris, A.G. & Kershaw, L.J. 1997. *Wetland Plants of Ontario.* Lone Pine Publishing, Edmonton, Vancouver, Redmond. 240 pp.

Otis, G.W. 1994. *Butterflies of Algonquin Provincial Park.* The Friends of Algonquin Park, Whitney, Ontario. 40 pp.

Quinn, N. 2002. *Algonquin Wildlife: Lessons in Survival.* Natural Heritage Books, Toronto. 246 pp.

Quinn, N.W.S. 2005. Reconstructing changes in abundance of White-tailed Deer, *Odocoileus virginianus*, Moose, *Alces alces*, and Beaver, *Castor canadensis*, in Algonquin Park, Ontario, 1860–2004. *Canadian Field-Naturalist* 119: 330–342.

Rawinski, T.J. 2008. Impact of White-tailed Deer overabundance in forest ecosystems: An overview. *Northern Area State and Private Forestry*, Forest Service, US Department of Agriculture. 8 pp.

Reader's Digest. 1982. *North American Wildlife.* Pleasantville, New York, Montreal. 576 pp.

Reid, F.A. 2006. *A Field Guide to Mammals of North America north of Mexico.* Peterson Field Guide Series. Houghton Mifflin. Boston, New York. 579 pp.

Rodd, T., ed. 2001. *The Firefly Encyclopedia of Trees and Shrubs.* Firefly Books, Willowdale, Ontario. 928 pp.

Rooney, T.P. 2001. Deer impacts on forest ecosystems: a North American perspective. *Forestry* 74: 201–208.

Rooney, T.P. & Waller, D. M. 2003. Direct and indirect effects of white-tailed deer in forest ecosystems. *Forest Ecology and Management* 181: 165–176.

Runtz, M.W.P. 1992. *Algonquin Seasons: A Natural History of Algonquin Park.* Stoddart Publishing, Toronto. 111 pp.

Runtz, M. 2008. *The Explorer's Guide to Algonquin Park.* Boston Mill Press, Erin, Ontario. 223 pp.

Rutter, R.J. & Strickland, D. 2002. *The Raven talks about Wolves.* Essays on wolves from Algonquin Park's popular newsletter, The Raven, 1963–2001. The Friends of Algonquin Park, Whitney, Ontario. 64 pp.

Rutter, R.J. & Strickland, D. 2003. *The Raven talks about Deer and Moose.* Essays on wolves from Algonquin Park's popular newsletter, The Raven, 1960–2002. The Friends of Algonquin Park, Whitney, Ontario. 60 pp.

Rutter, R.J. & Strickland, D. 2004. *The Raven talks about Fish and Lakes.* Essays on wolves from Algonquin Park's popular newsletter, The Raven, 1960–2003. The Friends of Algonquin Park, Whitney, Ontario. 56 pp.

Schneider, D. 2005. The Wild Ones. Ontario's Dogs and Cats. Ontario Nature, Federation of Ontario Naturalists, Toronto. *ON Nature* Vol. 44 (4): 38–40.

Sherman, J. 1990. *Indian Tribes of North America.* Todtri Productions Limited, New York. 144 pp.

Simard, J.H. 2001. *Habitat selection, ecological energetics, and the effects of changes in White Pine forests on breeding Red Crossbills (Loxia corvirostra) in Algonquin Provincial Park, Ontario.* M. Sc. thesis, McGill University. 130 pp.

Snow, D.W. & Perrins, C.M. 1998. *The Birds of the Western Palearctic.* Concise Edition. Oxford University Press, Oxford, New York. 1697 pp.

Strickland, D. 1992. *Bat Lake Trail. Basic Algonquin Ecology.* The Friends of Algonquin Park, Whitney, Ontario. 14 pp.

# BIBLIOGRAPHY

Strickland, D. 1993. *Berm Lake Trail. Algonquin Pine Forest Ecology* Bulletin No. 9. The Friends of Algonquin Park, Whitney, Ontario. 29 pp.

Strickland, D. 1994. *Booth's Rock Trail. Man and the Algonquin Environment.* The Friends of Algonquin Park, Whitney, Ontario. 14 pp.

Strickland, D. 1996. *Trees of Algonquin Provincial Park.* The Friends of Algonquin Park, Whitney, Ontario. 40 pp.

Strickland, D. 1997. *Track and Tower Trail. A Look Into Algonquin's Past.* The Friends of Algonquin Park, Whitney, Ontario. 14 pp.

Strickland, D. 1997. *Whiskey Rapids Trail. Algonquin River Ecology.* The Friends of Algonquin Park, Whitney, Ontario. 14 pp.

Strickland, D. 1998. *Beaver Pond Trail. Algonquin Beaver Ecology.* The Friends of Algonquin Park, Whitney, Ontario. 14 pp.

Strickland, D. 1998. *Brent Crater Trail. History of the Crater.* The Friends of Algonquin Park, Whitney, Ontario. 14 pp.

Strickland, D. 1999. *Mizzy Lake Trail. Wildlife in Algonquin.* The Friends of Algonquin Park, Whitney, Ontario. 14 pp.

Strickland, D. 1999. *Two Rivers Trail. Changes in the Algonquin Forests.* The Friends of Algonquin Park, Whitney, Ontario. 14 pp.

Strickland, D. 2000. *Centennial Ridges Trail. Historic Figures of Algonquin.* The Friends of Algonquin Park, Whitney, Ontario. 14 pp.

Strickland, D. 2000. *Hardwood Lookout Trail. Algonquin Hardwood Forest Ecology.* The Friends of Algonquin Park, Whitney, Ontario. 14 pp.

Strickland, D. 2000. *Hemlock Bluff Trail. Research in Algonquin.* The Friends of Algonquin Park, Whitney, Ontario. 14 pp.

Strickland, D. 2000. *Lookout Trail. Algonquin Geology.* The Friends of Algonquin Park, Whitney, Ontario. 14 pp.

Strickland, D. 2000. *Peck Lake Trail. Ecology of an Algonquin Lake.* The Friends of Algonquin Park, Whitney, Ontario. 18 pp.

Strickland, D. 2001. *Big Pines Trail. Ecology and History of White Pines in Algonquin.* The Friends of Algonquin Park, Whitney, Ontario. 14 pp.

Strickland, D. 2002. *Barron Canyon Trail. History of the Canyon.* The Friends of Algonquin Park, Whitney, Ontario. 10 pp.

Strickland, D. 2009. *Birds of Algonquin Provincial Park.* The Friends of Algonquin Park, Whitney, Ontario. 40 pp.

Strickland, D. 2009. What originally prevented, and what later permitted, the great northern expansion of White-tailed Deer? *Occasional Papers from Oxtongue Lake,* 1 (1): 1–40.

Strickland, D. *The Algonquin Park Book.* In press.

Strickland, D. & LeVay, J. 1998. *Wildflowers of Algonquin Provincial Park.* The Friends of Algonquin Park, Whitney, Ontario. 32 pp.

Strickland, D. & Ouellet, H. 1993. Grey Jay. In *The Birds of North America*, No. 40 (Poole, A., Stettenheim, P. & Gill, F., eds.). The Academy of Natural Sciences, Philadelphia, and The American Ornithologists' Union, Washington, D.C. 24 pp.

Strickland, D., Rutter, R. & Burke, P. 1993. *The Best of the Raven. 150 essays from Algonquin Park's popular newsletter in celebration of the Park Centennial 1893–1993.* The Friends of Algonquin Park, Whitney, Ontario. 220 pp.

Strickland, D. & Rutter, R. J. 2002. *Mammals of Algonquin Provincial Park.* The Friends of Algonquin Park, Whitney, Ontario. 48 pp.

Tanner, O. 1977. *The Canadians.* Time-Life Books, Chicago. 240 pp.

Tarvin, K.A. & Woolfenden, G.E. 1999. Blue Jay. In *The Birds of North America*, No. 469 (Poole, A. & Gill, F., eds.) The Birds of North America, Inc., Philadelphia. 32 pp.

Theberge, J.B. 1998. *Wolf Country: Eleven years of tracking the Algonquin wolves.* McClelland and Stewart, Toronto. 306 pp.

Theberge, J.B., Oosenburg, S.M. & Pimlott, D.H. 1978. Site and seasonal variation in foods of wolves, Algonquin Park, Ontario. *Canadian Field-Naturalist* 92: 91–94.

Theberge, J.B. & Strickland, D.R. 1978. Changes in wolf numbers, Algonquin Provincial Park, Ontario. *Canadian Field-Naturalist* 92: 395–398.

Thomson, S.C. 1974. Sight Record of a Cougar in Northern Ontario. *Canadian Field-Naturalist* 88: 87.

Thorn, R.G. 1991. *Mushrooms of Algonquin Provincial Park.* The Friends of Algonquin Park, Whitney, Ontario. 32 pp.

Tozer, R. 1997. *Checklist and Seasonal Status of the Birds of Algonquin Provincial Park.* Algonquin Park Technical

Tozer, R. 2012. *Birds of Algonquin Park.* The Friends of Algonquin Park, Whitney, Ontario.

Tozer, R. & Checko, N. 1996. *Algonquin Provincial Park Bibliography.* Algonquin Park Technical Bulletin No.12. The Friends of Algonquin Park, Whitney, Ontario. 116 pp.

Van Sickle, W. 1999. *Algonquin Park Visitor's Guide.* Stonecutter Press, Elmira, Ontario. 200 pp.

Voigt, D.R., Kolenosky, G.B. & Pimlott, D.H. 1976. Changes in summer foods of wolves in central Ontario. *Journal of Wildlife Management* 40: 663–668.

Wake, W.C., ed. 1997. *A Nature Guide to Ontario.* University of Toronto Press, Toronto.

Wilson, D. & Mittermeier, R.A., eds. 2009. *Handbook of the Mammals of the World.* Vol. 1. Carnivores. Lynx Edicions, Barcelona. 727 pp.

Wilson, D. & Mittermeier, R.A., eds. 2011. *Handbook of the Mammals of the World.* Vol. 2. Hoofed Mammals. Lynx Edicions, Barcelona. 885 pp.

Wood, D. 1995. *Wolves.* Smithbooks, Toronto. 109 pp.

Wright, B.S. 1972. *The Eastern Panther: A Question of Survival.* Clarke, Irwin & Co, Toronto. 180 pp.

*Black Bear foraging in Algonquin woods before the arrival of the long and cold winter. Bears are very shy and not easy to observe. If they come across humans they usually disappear in a hurry.*

# INDEX

**A**

*Acer rubrum* 167
*Acer saccharinum* 167
*Acer saccharum* 166
*Accipiter gentilis* 198
*Accipiter striatus* 64
*Aegolius acadicus* 213
*Achillea millefolium* 135
*Agelaius phoeniceus* 56
*Aix sponsa* 107
*Alces alces* 50
Algonquian, Algonquian people, Algonquins, Algonquin people 24, 25, 26, 27, 110–111
Algonquin Highlands 92, 110, 114, 166
Algonquin National Park 26
*Ameiurus nebulosus* 113
Amikeus Lake 42, 118–119, 139
*Anas platyrhynchos* 104
*Anas rubripes* 104
*Anax junius* 79
*Aquila chrysaetos* 196
*Ardea herodias* 102
Aspen, Largetooth 158, 160
Aspen, Trembling 158, 160
*Aythya collaris* 104

**B**

Barron Canyon 92, 94, 95. 96, 110
Barron River 92, 110, 167
Bass, Smallmouth 114, 117
Basswood 166
Bear, Black, bear 52, 55, 69, 76, 129, 146, 148, 154, 155, 156, 208-209
Beaver, 33, 42, 43, 44, 81, 160, 186, 193
Beech, American 58, 166
Berm Lake Trail 159
*Betula alleghaniensis* 177

*Betula papyrifera* 177
Big Crow Lake 130
Big Pines Trail 130
Birch, White 4, 122, 161, 168–169, 177
Birch, Yellow 58, 123, 139, 166, 177
Blackberry, Common 156
Blackbird, Red-winged 56
Blue Flag 82
Blue Vetch 134
Blueweed 134
Bobcat 173
Bolete, Birch 129
Bolete, Bitter 128
Bolete, King 128
*Boletus edulis* 128
*Bonasa umbellus* 66
Bonnechere River 110
*Branta canadensis* 105
Brown-eyed Susan 135
Bullhead, Brown 113
Bunchberry 67, 69
*Buteo jamaicensis* 62
*Buteo platypterus* 63
Butter-and-eggs 135

**C**

Canada Goldenrod 86
Canada Mayflower 68, 128
Canada Yew 139
*Canis lycaon* 175
*Castor canadensis* 42
*Cantharellus cibarius* 129
*Cathartes aura* 97
*Catostomus commersoni* 113
Cedar, Eastern White 125, 139
Centennial Ridges 10–11, 151
Centennial Ridges Trail 66
*Certhia americana* 150
Chalk-fronted Corporal 78
Cherry, Black 58, 166
Chickadee, Black-capped 188
Chickadee, Boreal 188
Chipmunk, Eastern, chipmunk 68, 69, 70, 71, 76, 119, 146, 171
*Chrysanthemum leucanthemum* 137

*Chrysemis picta* 81
Cisco 115
*Cladonia rangiferina* 158
*Claytonia caroliniana* 40
*Colaptes auratus* 37
Common Green Darner 78
*Cornus canadensis* 67, 69
*Corvus brachyrhynchos* 90
*Corvus corax* 96
Costello Creek 8-9, 65, 102
Cougar, Eastern 173
Creek Chub 112
Creeper, Brown 150
Crossbill, Red 190
Crossbill, White-winged 122, 190-191
Crow, American 90, 96
*Cyanocitta cristata* 153
*Cypripedium acaule* 68

**D**

*Danaus plexippus* 89
Deer, White-tailed, deer 6, 39, 53, 69, 124, 125, 138, 139, 140, 141, 142–143, 148, 155, 160, 173, 175, 177, 186, 196, 202
Dragonhunter 78, 79
Duck, American Black 104
Duck, Ring-necked 104
Duck, Wood 106, 107

**E**

Eagle, Bald 46, 196–197
Eagle, Golden 196
Eastern News 127
east side 92, 93, 94, 95, 112, 116, 131, 132, 149, 190
*Echium vulgare* 134
*Epilobium, angustifolium* 86
*Erythronium americanum* 41
*Esox lucius* 116
*Esox masquinongy* 112
*Eupatorium maculatum* 87

**F**

*Falcipennis canadensis* 34

# INDEX

*Falco columbarius*  64
Falcon, Peregrine  99
*Falco peregrinus*  99
Fallfish  112
*Felis concolor*  173
Fir, Balsam  122
Fireweed  86
Fisher  186, 193
Flicker, Northern  37
Flowery Blewits  126
Fossmill Outlet  92
Fox, Red, fox  69, 170, 171, 193, 194–195
Frog, Green  81

## G

*Gavia immer*  46
Golden Chanterelle  129
Goose, Canada  28–29, 105
Goshawk, Northern  198
Grackle, Common  90
Grosbeak, Evening  190
Grouse, Ruffed  32, 66, 67, 68, 198
Grouse, Spruce  6, 34, 35

## H

*Hagenius brevistylus*  79
*Haliaeetus leucocephalus*  196
Hardwood Lookout Trail  30–31, 38, 41, 123, 165
Hare, Snowshoe  172, 198
Hawk, Broad-winged  63, 146
Hawk, Red-tailed  62
Hawk, Sharp-shinned  64
Heal All  87
Hemlock, Eastern  123, 124, 139, 166
Heron, Great Blue  102, 163
*Hesperiphona vespertina*  190
Highway 60  16–17, 24, 28, 50, 59, 92, 94, 117, 135, 133, 148, 165
*Hirundo rustica*  94
Hummingbird, Ruby-throated  36

## I

Indian Pipe  126
*Iris versicolor*  82

## J

Jay, Blue  153
Jay, Grey  6, 146, 153, 186, 198
Junco, Dark-eyed  188
*Junco hyemalis*  188
Juniper, Common  93
*Juniperus communis*  93

## K

Kingfisher, Belted, kingfisher  103, 180

## L

*Ladona julia*  78
Ladyslipper, Pink  68
Lake Herring  115
Lake of Two Rivers  202-203
Lake Travers  112
Larch, American, Eastern  161
*Larix laricina*  161
*Leccinum scabrum*  129
*Lepista irina*  126
*Lepomis gibbosus*  117
*Libellula pulchella*  79
lichens  158, 159
Lily, Bullhead  79, 81
Lily, White Water  81, 83
Lily, Yellow Pond  81
Lily-of-the-Valley, Wild  68
*Limenitis arthemis*  88
*Linaria vulgaris*  135
*Lontra canadensis*  202
Lookout Trail, cliff  4, 158
Loon, Common, loon  32, 46, 47, 48–49, 146, 180
Loosestrife, Purple  26–27, 134
*Lophodytes cucullatus*  181
*Loxia curvirostra*  190
*Loxia leucoptera*  190–191
*Luxilus cornutus*  181
Lynx  172, 173
*Lynx canadensis*  172
*Lynx rufus*  173
*Lysimachia terrestris*  86
*Lythrum salicaria*  134

## M

*Maianthemum canadense*  68
Mallard  104, 123
Maple, Red  58, 146, 166, 167, 168–169
Maple, Silver  167
Maple, Sugar  32, 33, 58, 59, 92, 138, 146, 165, 166
Marten, American, marten  119, 193
*Martes americana*  119
*Megaceryle alcyon*  103
Merganser, Common  108–109, 180, 181
Merganser, Hooded  181
*Mergus merganser*  180
Merlin  64, 65
Mew Lake  79, 184–185
*Micropterus dolomieui*  117
Mink  193
Mizzy Lake Trail  124, 154, 178–179
Moccasin Flower  68
Monarch  89
*Monotropa uniflora*  126
Moose  6, 7, 20, 27, 33, 50, 51, 52, 53, 54–55, 69, 76, 81, 139, 154, 155, 160, 161, 175, 186, 196, 202
Moss, hair-cap  158
Mouse, Deer  69, 165
Muskellunge  110, 112

## N

*Notemigorus crysoleucas*  181
*Notophthalmus viridescens*  127
*Notropis heterolepis*  181
*Nuphar variegata*  79, 81
Nuthatch, Red-breasted  125, 150
Nuthatch, White-breasted  150
*Nymphaea odorata*  81

## O

Oak, Bur  70
Oak, Red  70, 93, 148, 149
*Odocoileus virginianus*  138
Opeongo Lake  47, 110–111, 114
Opeongo River  116
*Oreothlypis ruficapilla*  61

211

# ALGONQUIN PARK: A PORTRAIT

Osprey  64, 180
Ottawa River  25, 26, 92, 110
Otter, River  202
Owl, Barred  189
Owl, Northern Saw-whet  213
Ox-eye Daisy  136–137
Oxtongue River  84–85, 90–91, 108–109, 110, 167

**P**

Painted Suillus  128
Painted Turtle  81
Paleo-indians  25
*Pandion haliaetus*  64
Peck Lake  152–153, 156–157
*Perca flavescens*  117
Perch, Yellow  117
*Perisoreus canadensis*  153
*Peromyscus maniculatus*  165
Petawawa River  110, 112, 167
Phoebe, Eastern  94, 95
*Picea glauca*  122
*Picea mariana*  122
*Picea rubens*  123
Pickerelweed  83
*Picoides pubescens*  188
*Picoides villosus*  36
Pike, Northern  114, 116
Pine, Eastern White  27, 59, 92, 93, 96, 120, 130, 131, 132, 166, 190
Pine, Jack  92, 132, 133
Pine, Red  92, 93, 131, 132
*Pinus banksiana*  132
*Pinus resinosa*  131
*Pinus strobus*  130
*Poecile atricapillus*  188
*Polygonia interrogationis*  88
*Pontederia cordata*  83
Poplar, Balsam  160
*Populus balsamifera*  160
*Populus grandidentata*  160
*Populus tremuloides*  160
*Procyon lotor*  72
*Prunella vulgaris*  87
Pumpkinseed  117

**Q**

*Quercus macrocarpa*  70
*Quercus rubra*  70, 93, 148
Question Mark  88
*Quiscalus quiscula*  90

**R**

Raccoon, raccoon  69, 72, 149
*Rana clamitans*  81
Raspberry, Red  156
Raven, Common  90, 96
Red Spruce Side Trail  123
Ring-neck Pond  16–17
Robin, American  127
*Rubus allegheniensis*  156
*Rudbeckia hirta*  135
*Russula claroflava*  129
Russula, Yellow Swamp  129
Rutter Lake  74-75

**S**

*Salvenius fontinalis*  114, 182
*Salvenius namaycush*  115
Sapsucker, Yellow-bellied  4, 36, 150
*Sayornis phoebe*  95
*Scolopax minor*  178
*Semotilus atromaculatus*  112
*Semotilus corporalis*  112
*Setophaga caerulescens*  60
*Setophaga coronata*  20-21, 33, 61
*Setophaga fusca*  124
*Setophaga pensylvanica*  61
Shiner, Blacknose  181
Shiner, Common  181
Shiner, Golden  181
*Sitta canadensis*  150
*Sitta carolinensis*  150
*Solidago canadensis*  86
*Sphyrapicus varius*  36
Splake  115, 117
Spotted Joe-pye Weed  87
Spring Beauty, Northern  32, 39, 40
Spruce Bog  162-163
Spruce Bog Boardwalk  122
Spruce Bog Trail  12-13, 35

Spruce, Black  63, 120-121, 122, 123, 190
Spruce, Red  35, 123
Spruce, White  122, 190
Squirrel, Red  120, 122
*Strix varia*  189
*Suillus pictus*  128
Sunfish  117
Swallow, Barn  94, 95
Swamp Candle  86, 87

**T**

Tamarack  146, 161, 190
*Tamiasciurus hudsonicus*  120
*Tamias striatus*  71
Tea Lake  180
*Thuja occidentalis*  125
*Trillium erectum*  39
*Trillium grandiflorum*  39
Trillium, Painted  32, 39
Trillium, Red  32, 39
*Trillium undulatum*  39
Trillium, White  39, 138
Trout, Brook  41, 114, 115, 116, 117, 182–183
Trout, Lake  115, 116
Trout Lily  32, 41
*Tsuga canadensis*  124
*Turdus migratorius*  127
Twelve-spotted Skimmer  78, 79

**U**

*Ursus americanus*  154

**V**

*Vicia cracca*  134
*Viola sororia*  40
Violet, Sweet  40
Violet, Wooly Blue  40
Viper's Bugloss  134
Visitor Centre  14–15, 20, 33, 122
*Vulpes vulpes*  170
Vulture, Turkey  97

# INDEX

**W**
Walleye   114
Warbler, Blackburnian   124
Warbler, Black-throated Blue   60
Warbler, Chestnut-sided   61
Warbler, Myrtle   61
Warbler, Nashville   61
Warbler, Yellow-rumped   20–21, 33, 61
West Rose Lake   2-3, 120–121
White Admiral   88, 89
Whitefish   115
Whitefish Lake   10–11, 151
White Sucker   113
Wolf, Eastern, wolf, wolves   6, 44, 52, 55, 58, 76, 96, 140, 141, 147, 174, 175, 186, 196, 200–201, 202
Wolf, Grey   175
Wolf Howl Pond   6, 178–179
Wolf, Red   175
Woodcock, American   178
Woodpecker, Downy   188
Woodpecker, Hairy   36, 188
Woodpecker, Pileated   107

**Y**
Yarrow   135
Yellow Toadflax   135
York River   110, 112

**Above:** *The **Northern Saw-whet Owl**, Aegolius acadicus, is a spring and summer resident and breeding species in Algonquin Park. Its repeated monotonous low whistles on early spring nights announce the beginning of their breeding season.*